A
Course in
CRYSTALLOGRAPHY

JAK Tareen
TRN Kutty

Universities Press (India) Limited

Registered Office
3-5-819 Hyderguda, Hyderabad 500 029 (A.P.), India

Distributed by
Orient Longman Limited

Registered Office
3-6-272 Himayatnagar, Hyderabad 500 029 (A.P.), India

Other Offices
Bangalore / Bhopal / Bhubaneshwar / Chandigarh / Chennai
Ernakulam / Guwahati / Hyderabad / Jaipur / Kolkata
Lucknow / Mumbai / New Delhi / Patna

© Universities Press (India) Limited 2001

First published 2001
ISBN 81 7371 360 X

Typeset by
Venture Graphics
Chennai 600 031

Printed at
Orion Printers
Hyderabad 500 004

Published by
Universities Press (India) Limited
3-5-819 Hyderguda, Hyderabad 500 029

PREFACE

The stimulus to write a book covering the basics of crystallography emerged from the constant interaction we have had with our students for over 30 years. Crystallography is one of those branches of science, the rudimentary knowledge of which is essential to beginners in physics, chemistry, geology, metallurgy, molecular biology and specifically to materials science and mineralogy. Owing to the interdisciplinary nature of the subject, introductory books covering the broader aspects of crystallography and catering to the needs of students with different backgrounds are not commonly available, that too with a logical and sequential development of the topic. Exhaustive treatises on crystallography by well-known authors are rather difficult to comprehend, because of the special ways of presentation of the subject depending upon individual perceptions and preferences. Most students shy away from crystallography because of the difficulties in three-dimensional visualization required to understand symmetry operations, crystal geometry relations, space lattices, symmetry related physical properties, etc. Simplicity of explanations, minimal usage of mathematical expressions and the brevity of the text with appropriate illustrations are often sought by students in crystallography classes. Keeping these points and the prescribed syllabi of most of the Indian Universities in view and the benefits from our own classroom experiences, we tried to develop the subject in this book so as to have a continuous reading from the elementary levels and to demand no background prerequisites.

The book is divided into six chapters. The first chapter presents the concepts of symmetry, external symmetry operations, the recognition of point groups and their restriction in number, based on the permissible combinations of symmetry elements. The second chapter takes the reader into the physical description of crystals and makes use of the stereographic projections to present the crystals on the two-dimensional planes. The third chapter introduces the internal (structural) symmetry in crystals through the concepts of translational operations and the space lattice. A reader, by now, ought to be conversant with the Bravais lattices, simple space groups, etc and understand how the external symmetry operations of rotation, reflection and inversion develop into more complex screw axes and glide planes in the internal structure, thereby extending the knowledge to 230 space groups. The fourth chapter probes into the intricacies of the internal arrangements in crystals by way of the concepts in close packing of hard spheres and the nature and geometry of crystal lattices leading to the common structural types encountered in inorganic materials and minerals. The fifth chapter introduces the common physical properties, particularly the crystal optics, used for the characterization of crystalline solids. The most powerful tools for investigating the internal arrangements of atomic and molecular constituents in crystals are the diffraction techniques, making use of the appropriate radiations. The last chapter presents the basic principles of X-ray, electron and neutron diffraction by crystals and the interpretation of the diffraction patterns. We have given the bibliography of some selected books for further reading, at the end of this book.

We have sought to present a comprehensive basic course in crystallography to Indian students. We believe that teachers will find this book equally useful as a basic source of information. There is considerable scope for improvements and we invite suggestions and reorganizational comments from other experienced teachers of crystallography. The same invitation holds good for the alert students, as well.

We hope that these suggestions will find their place in the future edition.

J A K Tareen   **T R N Kutty**
Mysore   Bangalore

CONTENTS

CHAPTER 1: EXTERNAL SYMMETRY

CHAPTER 2: DESCRIPTION OF CRYSTALS

CHAPTER 3:   INTERNAL SYMMETRY AND CRYSTAL LATTICE

CHAPTER 4: STRUCTURAL PRINCIPLES IN CRYSTALS

CHAPTER 5: THE PHYSICAL PROPERTIES OF CRYSTALS

CHAPTER 6:   DIFFRACTION BY CRYSTALS—X-RAY CRYSTALLOGRAPHY

External Symmetry

1.1 INTRODUCTION

Among the inanimate objects, crystals have attracted man from time immemorial. The enchanting colours, the smooth surfaces with scintillating reflections of light, the definite and varied shapes with sharp edges, the deep transparency of some perfect crystals, all together aroused the aesthetic sense of the early man who used them as ornaments. Parallel with the growth of other sciences, through time, grew the curiosity of mankind to understand, more quantitatively the properties of crystals. The utility of crystals has been extended from the bounds of ornaments to several useful applications in optical, electrical and opto-electronic devices. The fantasy of their external beauty was understood more thoroughly through the natural laws of mathematics, physics and chemistry. The contents of the crystals and their 'insides' were explored, analysed and understood by modern methods of diffraction as well as with the help of spectroscopic techniques. The external shapes, planes and colours were correlated with the internal atomic content and their arrangements in unequivocal terms. Thus grew a science: the study of crystals—Crystallography.

1.1.1 What is a Crystal?

The word crystal originates from the Greek word *Krystallos* meaning clear transparent ice. In the Middle Ages, this word was extended to include quartz or rock crystals which were believed to have been formed by the intense freezing of water on the Alps mountains into a permanent form of ice. To a present day student of Crystallography, the crystal is a chemical compound in the shape of a solid polyhedron bounded by definite planes. Their shape and symmetry is a manifestation of the internal atomic arrangement extending in three dimensions, in an orderly way.

1.1.2 Crystal Elements

The intersection of two smooth faces (forms) of a crystal make an edge. The angle between the normals to these faces is called the interfacial angle which remains constant, irrespective of the size of the crystal. The intersection of more than two faces makes a corner, called the solid angle. The face, the edge and the solid angle are the three elements of a crystal.

1.2 DEFINITION OF SYMMETRY

The human eye has the instant ability of judging the symmetry of an object and comparing the same between different objects of varying shapes. Our instinct or judgement suggests that the planar objects in Fig. 1.1 are arranged in the order of decreasing symmetry. If any part of the object is identical to an other part, we consider it as possessing some symmetry. In other words, the number of times the original position can be interchanged without being able to distinguish the new position, will, to some extent, quantify the symmetry of an object. A square can be interchanged this way four times, an equilateral triangle three times, a rectangle twice and an irregular planar object only once. This quantitatively defines the degree of decreasing symmetry of these planar objects. This systematic repetition of identical features is characteristic of a symmetrical object. It has been considered by crystallographers that the external symmetry arises because of the symmetry of the internal arrangement of atoms or molecules or the building blocks, generally known as unit cells. The concept of unit cell will be developed in Chapter 3.

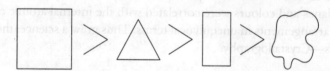

Fig. 1.1 Planar objects in the order of decreasing symmetry

1.3 SYMMETRY ELEMENTS

The quantification or the measurement of symmetry of an object is possible only when one examines the appearance of the identical and interchangeable parts with reference to:

a) a point at the centre of the object,
b) an imaginary line passing through the centre of the object, and
c) an imaginary plane passing through the centre of the object.

1.3.1 Symmetry with Reference to a Point

If identical and interchangeable parts appear symmetrically equidistant on either side of the point in the object, a centre of symmetry is said to exist. In other words, if any part of an object is joined by a line to this point in the object and extended further, the line will meet an identical part at the same distance away from the chosen point. One can identify this symmetry by inverting the object, and the operation involved is called an 'inversion' (Fig. 1.2). Thus, the object is said to have an inversion centre. Alternatively, the object is said to be centrosymmetric.

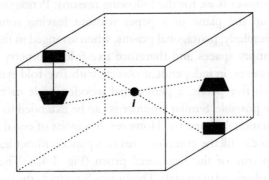

Fig. 1.2 Symmetry with reference to centre *i* which is the geometric centre of a room. Any two diametrically opposite points in this room are equidistant from this central point

1.3.2 Symmetry with Reference to a Line

If the interchangeable parts of an object appear around a line or an axis, then the object is said to possess an axis of symmetry. The number of times such superimposable positions can be obtained around an axis defines the degree of the rotational symmetry. In other words, if the object is rotated around an imaginary axis passing through it, by an angle of 360°, the interchangeable position can appear once, twice, thrice, four times and so on, at a repeat angle of 360°, 180°, 120°, 90°, and so on, respectively. When the repeat angle is $360/X$, where X is 1, 2, 3 and so on, the number of times the interchangeable parts appear in one complete rotation is X. This operation currently performed is called a 'rotation'. These are designated as 1-, 2-, 3-, 4- and so on-fold axis (eg. four-fold). Observe the effect of the rotational operation on the appearance of the object in Fig. 1.3. In all the cases, the object appears the same, as though nothing has happened and after five such rotations of 90° each, the object will be back to its original position which is called the identity position.

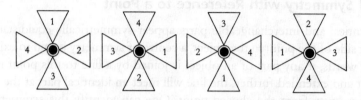

Fig. 1.3   Four-fold symmetry. The changed petal positions through relation by 90° can not be distinguished unless the petals are numbered

Can the rotational symmetry axis be of unrestricted value, i.e., say X of any value? The answer is *no*, for the following reasons: Pentagons of equal sides cannot fill up the plane of a paper without leaving some empty space (Fig. 1.4a). Similarly, pentagonal prisms, when arranged in three dimensions, also leave empty spaces and therefore five-fold symmetry cannot exist in crystals. However, in independent objects with five-fold symmetry, such as a flower with five petals, a pentagonal wooden table or even a five-fold molecule are possible. Similar arguments can be extended to 7-, 8-, or higher orders of rotational symmetry. However, hexagons of equal size arranged in juxtaposition can fill the complete area of a paper without leaving any space. The same is true of the hexagonal prism (Fig. 1.4b). Therefore, six-fold symmetry is observed in crystals. This uniquely restricts the possible rotational symmetry in crystals to 1-, 2-, 3-, 4- and 6- fold only and proof to this effect is given in Chapter 3. This is how the crystals differ from other symmetrical objects around us.

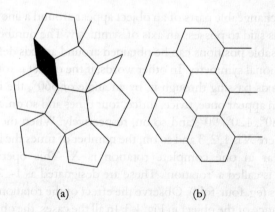

(a)   (b)

Fig. 1.4   (a) Pentagons packed in a plane leave empty spaces, whereas (b) hexagons leave no space and hence a five-fold symmetry is absent in crystalline solids

1.3.3 Symmetry with Reference to a Plane

The third way of evaluating symmetry, is with reference to an imaginary plane passing through an object. The two divided parts are related to one another like images in the mirror. Imagine that the object is cut along the plane and one half is removed. If a mirror is placed on the plane (along the cut surface) then the remaining half of the object would be sufficient to complete the object because of the reflected image (Fig. 1.5). The symmetry operation involved here is called 'reflection'.

There are two types of mirror planes possible (Figs. 1.6a & b). One type of mirror plane is parallel to two of the three Cartesian axes, x, y or z *axes* of the object. The second type is parallel only to one of these axes.

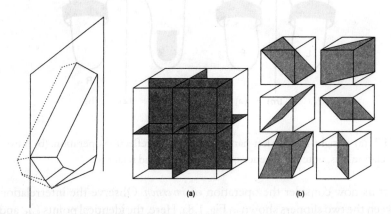

(a)   (b)

Fig. 1.5 Symmetry with reference to a mirror plane where one half is the mirror image of the other

Fig. 1.6 Types of mirror planes: (a) Planes parallel to any two of the reference axes (Cartesian axes x, y or z) (b) Planes parallel to any one of the reference axes

1.3.4 Special Features of Symmetry Elements

The different operations namely, inversion, rotation and reflection are called *symmetry operations*. These are movements necessary to bring about points of coincidence with symmetrically equivalent positions, in order to identify the symmetry elements. One can visualize the movements only by considering an asymmetric object. Observe Fig. 1.7b where a slipper for the right foot is rotated around an axis by 180°, and this exposes the sole of the slipper. Even in the new position, the slipper is of the right foot. In contrast, see Fig. 1.7a where two slippers are related to one another by a mirror plane. This is because the movement of points are perpendicular to and equidistant from the plane. This means that the reflection or mirroring of the left slipper produced a right slipper. The same relation can be illustrated with a left-hand reflected in a mirror. This is known as the right-hand left-hand relationship (*enantiomorphism*) in crystals. This discussion brings out the point that rotation leaves the object without any change of hand and is called a symmetry operation of the "first kind".

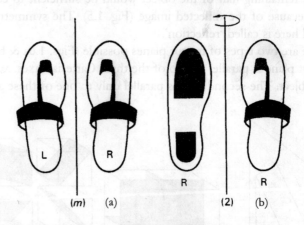

Fig. 1.7 (a) A right-hand to left-hand relation with a reflection operation. (b) A two-fold relation axis; note the absence of right- to left-hand relation

Let us now consider the operation of *inversion*. Observe the interrelation between the two slippers shown in Fig. 1.8a. Here, the identical points (a_1 and a_2 or b_1 and b_2) are equidistant across the point (i). The movements are not perpendicular to one another and the inverted footwear would fit onto a left foot. Therefore, the operations of both reflection and inversion involve the change from right to left, and these are called symmetry operations of the "second kind".

1.4   IMPROPER AXES

Besides the above three kinds of elements, there exist complex symmetry elements. These can be imagined as a combination of symmetry operations of the first kind with any one of the second kind. Thus, we can have a combination of: (a) rotation axis with reflection—the *rotoreflection* axis, (b) rotation axis with inversion centre—the *rotoinversion axis*. The resultant symmetry elements are called *improper axes*, whereas the simple rotation axes of 1-, 2-, 3-, 4- and 6-fold symmetry are called the *proper axes*. Since the rotoreflection and rotoinversion operations produce repetitive symmetry properties, only one of them is sufficient to describe the improper axes of symmetry. Conventionally, the rotoinversion axes are most widely used by crystallographers. The rotoreflection axes are not further dealt with in this book.

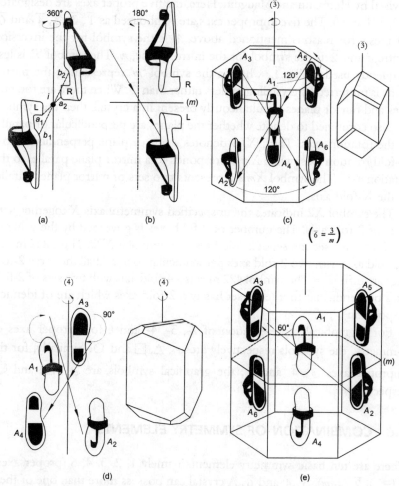

Fig. 1.8 The symmetry operations of the second type: (a) Rotation by 360° followed by the inversion ($\overline{1}$ axis), after this operation the right foot fits into the left foot. (b) Rotation by 180° followed by inversion ($\overline{2}$ axis). Here again the left-foot slipper fits into the right foot after the operation. The net effect is a reflection. (c) The operation of rotation by 120° and inversion beginning from position a_1 until we reach the starting positions ($\overline{3}$ axis). Here all the slippers in the upper portion are of the right foot and those in the lower portion are of the left foot. (d) Operation of rotation by 90° and inversion four times ($\overline{4}$ axis) (e) Operation of rotation by 60° followed by inversion six times results in a $\overline{6}$ axis. The symmetry produced is a 3-fold axis and a mirror plane normal to the axis. Hence $\overline{6} = 3/m$

1.5 SYMBOLS FOR THE SYMMETRY ELEMENTS

Although there are different methods used to represent the symmetry elements in short form, the internationally accepted symbols are those devised by Hermann and Mauguin. Here, the five proper axes are designated 1, 2, 3, 4 and 6. The five improper axes are symbolized as $\bar{1}$, $\bar{2}$, $\bar{3}$, $\bar{4}$ and $\bar{6}$. Of these, for reasons mentioned above, $\bar{1}$ is the symbol for the inversion centre, *i* and $\bar{2}$ is the symbol for the mirror plane, *m*. The symbol '*i*' is less frequently used unlike $\bar{1}$, whereas the symbol '*m*' representing the mirror plane is preferred by crystallographers rather than $\bar{2}$. When there are rotation axes and mirror planes simultaneously present in a crystal, then the symbols are also combined to denote whether the planes are perpendicular or parallel to the rotation axes. Thus, X/m denotes a mirror plane perpendicular to a X-fold rotation axis, while Xm corresponds to a mirror plane parallel to the rotation axis. The symbol Xmm represents two sets of mirror planes parallel to the X-fold axis.

The symbol $X2$ indicates the unspecified symmetry axis X together with a set of 2-fold axis. The number of 2-fold axes is governed by the value of X. If there is a second set of 2-fold axes, the symbol is $X22$. Thus, 222 means a 2-fold axis with two 2-fold axes perpendicular to it, and all the three 2-fold axes are different. The symbol 422 means a 4-fold axis with two sets of 2-fold axes perpendicular to it (each set has two 2-fold axes which are of identical disposition).

For a graphical representation of 2-, 3-, 4- and 6-fold proper axes of symmetry, the symbols respectively are $\bigcirc$, $\triangle$, $\square$ and $\bigcirc$. Similarly, for the improper axes $\bar{3}$, $\bar{4}$ and $\bar{6}$, the graphical symbols are $\blacktriangle$, $\ominus$ and $\oslash$, respectively.

1.6 COMBINATION OF SYMMETRY ELEMENTS

There are ten basic symmetry elements namely 1, 2, 3, 4, 6 (proper axes), $\bar{1}(= i)$, $\bar{2}$ $(= m)$, $\bar{3}$, $\bar{4}$ and $\bar{6}$. A crystal can possess more than one of these elements simultaneously. There may also exist multiples of the same kind of elements, but with certain restrictions on each combination of elements in a crystal which arise out of the following factors.

1.6.1 Non-interference in the Symmetry Characteristics of Individual Elements

In Fig. 1.9a, a 3-fold axis is combined with three 2-fold axes so that the symmetry characteristics of both the sets of axes co-exist simultaneously. On the other hand, if there were to be only one 2-fold axis combined with a 3-fold axis, the symmetry characteristics of both the rotation axes are lost. The reader may convince himself with drawings of his own.

Yet another illustration is shown in Fig. 1.9b, wherein a 3-fold axis is combined with three other 3-fold axes. Here although the 3-fold characteristics of element number 1 is preserved, the 3-foldness of the other three is lost. This type of combination is, therefore, impossible in crystals. This condition of non-interference of symmetry characteristics is to be borne in mind when any two or more symmetry elements are combined.

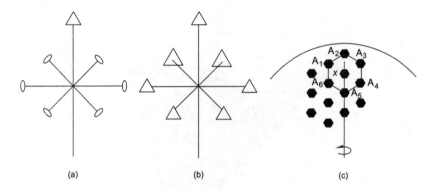

(a)   (b)   (c)

Fig. 1.9 A combination of rotation axis: (a) 3-fold axis and three perpendicular 2-fold axes is compatible. (b) A combination of perpendicular 3-fold axis with three more perpendicular 3-fold axes is incompatible. (c) Likewise a combination of more than one 6-fold axis leads to the generation of spherical symmetry, thus losing the 6-fold character.

1.6.2 A Combination of Symmetry Elements should not Result in a Spherical Symmetry which has Infinite-fold Symmetry

Fig. 1.9c illustrates a point where two 6-fold axes are combined at any random angle. In order to retain the 6-fold character of element 'A_1' we have to add five more 6-fold elements around it. Each of these 6-fold elements demand another set of six 6-fold elements around each of them. This process continues infinitely, resulting in a spherical symmetry which in turn, has infinite-fold axes.

1.6.3 The Intersection of Two Symmetry Axes Generates a Third Symmetry Axis

This statement is illustrated in Fig. 1.10, wherein a 4-fold axis (X-fold, in general) is combined with a 3-fold axis (Y-fold, in general) to generate a third 2-fold axis (Z-fold axis in general). The points A_1, A_2, A_3 and A_4 repeat at 90°

on a plane perpendicular to a 4-fold axis. Through rotation around a 3-fold axis at the corner, A_2 repeats at 120° at A_2' and A_1 and lies on the plane normal to this 3-fold axis. Now, in the presence of these two axes, the identical points (A_2, A_3, A_4, A_1, etc) lie on a diagonal plane, which incidentally is normal to the third symmetry axis which is an inevitable 2-fold axis generated. In general, if two symmetry axes X and Y are made to intersect they generate the third symmetry axis Z, each with its own symmetry characteristics and without interfering with one another as mentioned in Section 1.6.1. The compatibility of such combinations is discussed below.

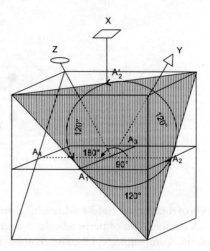

Fig. 1.10   Generation of symmetry axis when two symmetry axes combine. Here a combination of 4-fold and 3-fold axes generates a 2-fold (2) axis. There are four 3-fold (Y) axes emerging from each of the corners of the cube. There are four 2-fold axes (Z) emerging from the middle of the cube edges. This way the 4-fold character of the X-axis is preserved. [See Fig. 1.13(b)]

A_1 is rotated on a plane (unshaded) perpendicular to X-axis by 90° to A_2 (and A_3, A_4). When the same A_1 is rotated on a second axis Y, it repeats at A_2 and A_2' at angles of 120°, on a plane perpendicular to this 3-fold axis. This combination of X(4-fold) and Y(3-fold) symmetry axes automatically generates the third inevitable axis Z, which is perpendicular to the diagonal plane on which lie the identi-points A_2 and A_4, 180° apart.

1.6.4 Permissible Combinations of Symmetry Axes

The restriction involved in X-fold, Y-fold and Z-fold symmetry axes can be explained in terms of the value of X, Y and Z as well as interaxial angles $X^\wedge Y$, $X^\wedge Z$ and $Y^\wedge Z$. These can be established only through a mathematical relation involving spherical triangles (Fig. 1.11). Those readers who have no exposure to solid geometry and spherical trigonometry may skip this section, except for making use of the conclusions given in Table 1.1 and the corresponding illustration given in Fig. 1.12.

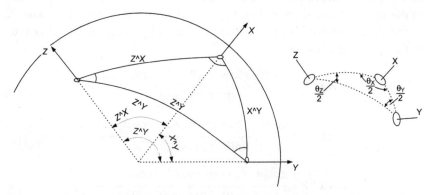

Fig. 1.11 The three intersecting symmetry axis of a sphere define a spherical triangle on the surface whose sides correspond to angles between axis (interaxial angles) and these angles are equal to one half the repeat angles, i.e. $\theta_{x/2}, \theta_{y/2}$ and $\theta_{z/2}$ for each symmetry axis X, Y & Z

The three intersecting axes X, Y and Z define a spherical triangle whose sides (arcs) are equal to the interaxial angles $X^\wedge Y$, $Y^\wedge Z$ and $X^\wedge Z$. The angles of the spherical triangle are $\theta_{x/2}$, $\theta_{y/2}$ and $\theta_{z/2}$, according to the constructions of Euler (Rodrigues construction). Each of these represent half the repeat angles associated with the X, Y and Z axes, respectively. According to the theory of the spherical triangle, the sum of the angles of a sphere should not be less than 180° and not more than 540°. This means,

$$180° < \theta_{x/2} + \theta_{y/2} + \theta_{z/2} < 540°. \qquad (1.1)$$

Since $\theta_x = 360°/X$, $\theta_y = 360°/Y$ and $\theta_z = 360°/Z$, the inequality expression (1.1) becomes

$$180° < 360°/2X + 360°/2Y + 360°/2Z < 540°$$

$$\text{i.e. } 1 < 1/X + 1/Y + 1/Z < 3. \qquad (1.2)$$

This inequality condition should be true for any compatible combination of X, Y and Z. To illustrate this point, let us consider the case of $X = 4$, $Y = 3$ and $Z = 2$ which means that $1/4 + 1/3 + 1/2 = 1.08$, which obeys the expression (1.2). Hence, 4, 3 and 2 is a compatible combination.

But consider a case of $X = 2$, $Y = 4$ and $Z = 6$, then $1/2 + 1/4 + 1/6 = 0.917$, and this does not obey the inequality condition. In the strict sense, X, Y and

Z cannot have 1-fold axis as it does not bring about an interchangeable position. However, there are five exceptional cases where X, Y and Z equals 444, 344, 334, 333 and 244 which intercept compatibly as in the case of a cube (which is explained later), and yet do not obey the inequality condition. The reason for the above exceptional combination arises from the compatibility of the angles of intersection. The angles at which the three symmetry axes X, Y and Z can compatibly intercept are evaluated by using the law of cosines in the spherical triangles.

$$\cos X^\wedge Y = \frac{\cos(\theta_z / 2) + \cos(\theta_x / 2) * \cos(\theta_y / 2)}{\sin(\theta_x / 2) * \sin(\theta_y / 2)} \tag{1.3}$$

$$\cos Y^\wedge Z = \frac{\cos(\theta_x / 2) + \cos(\theta_y / 2) * \cos(\theta_z / 2)}{\sin(\theta_y / 2) * \sin(\theta_z / 2)} \tag{1.4}$$

$$\cos X^\wedge Z = \frac{\cos(\theta_y / 2) + \cos(\theta_z / 2) * \cos(\theta_x / 2)}{\sin(\theta_z / 2) * \sin(\theta_x / 2)} \tag{1.5}$$

where X, Y and Z represent 2-, 3-, 4- or 6-fold (or corresponding $\overline{2}$ -, $\overline{3}$ -, $\overline{4}$ - or $\overline{6}$ -fold axes) the angles are $180°/2 = 90°$, $120°/2 = 60°$, $90°/2 = 45°$ or $60°/2 = 30°$, respectively.

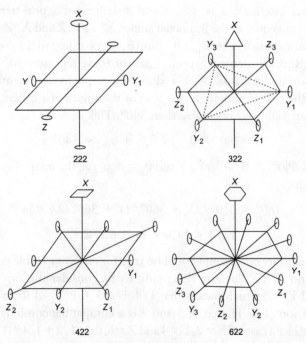

Fig. 1.12   The first four compatible combinations of symmetry axes from Table 1.1

The compatible values of X, Y and Z with the corresponding intersection angles are listed in Table 1.1. The angles of intersection are being calculated from Eqs. (1.3) to (1.5). The first four combination of axial symmetry are illustrated in Fig. 1.12. Y and Z represent the 2-fold axis, although in the strict sense both Y and Z constitute a single set of symmetrically related 2-fold axes. The last two combinations in Table 1.1 are illustrated in Fig.1.13. The angles $X \wedge Y$ equal 54°44′, which is the angle between the body diagonal and the perpendicular to a face of the imaginary cube (also called the 'guide cube'). The combination of a 4-fold (X) with 3-fold (Y) axis generates the 2-fold axis (Z), which corresponds to the line joining the mid-points of the opposite edges of the cube. In order to preserve the symmetry characteristic of 4, 3 and 2-fold axes, additional symmetry elements have to be produced. That means there will be three 4-fold, four 3-fold and six 2-fold axes, as shown

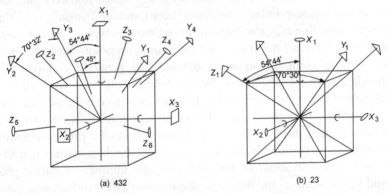

(a) 432   (b) 23

Fig. 1.13 The last two compatible combinations of the symmetry axis from Table 1.1. (a) To intersect compatibly, a 4-, 3-, and 2-fold axis must appear at the positions shown as X, Y and Z on a guide cube. (b) When a 4-fold axis X_1 in 'a' is replaced by 2-fold, then for the 2-fold and 3-fold axis to exist compatibly, the 3-fold axis should occur at Z_1 and Y_1. The X_2 and X_3 requires two more 2-fold axes and 3-fold axes. Hence, three 2-fold and four 3-fold axis only are compatible

Table 1.1 The three compatible symmetry axes and their intersection angles

X	Y	Z	$X \wedge Y$	$X \wedge Z$	$Y \wedge Z$
2	2	2	90°	90°	90°
3	2	2	90°	90°	60°
4	2	2	90°	90°	45°
6	2	2	90°	90°	30°
4	3	2	54°44′	45°	35°16′
2	3	3	54°44′	54°44′	70°32′

in Fig.1.13a, emerging from the mid-point of the cube. If the 2-fold axes take the place of three 4-fold axes (Fig. 1.13b), then the number of symmetry axes reduces, and only four 3-fold axes emerging from the body diagonals persist. The angles between any two 3-fold axes becomes 70°32′, whereas the angle between 2-fold and 3-fold remains at 54°44′. In Fig. 1.13b, the guide cube is purely imaginary and does not exist.

1.6.5 Permissible Combination of Symmetry Axes with Inversion Symmetry (Centre of Symmetry) or the Mirror Planes

When one combines an inversion centre with 1-, 2-, 3-, 4- or 6-fold proper axes, then the 2-, 4- and 6-fold axes (even axes) generate a mirror plane (Fig. 1.14) perpendicular to the rotation axis. These are represented by the symbols $2/m$, $4/m$ and $6/m$. The other two, i.e., 1- and 3-fold axes (odd axes), become the corresponding improper axes $\bar{1}$ and $\bar{3}$. Instead of a single symmetry axis, one may think about more than one type of proper axes, as discussed in Section 1.6.4, intersecting at a common point. Then each one of these axes in combination will independently generate a mirror plane perpendicular to itself. That means 2-, 4- or 6-fold axis involved in the combination will each become $2/m$, $4/m$ or $6/m$ respectively, whereas the 3-fold axis will become the $\bar{3}$ axis. The question of considering the 1-fold axis in a combination does not arise. The combination of a proper axis with a mirror plane can be done in two ways: a) parallel and b) perpendicular to the axis under consideration. In the first case, the number of mirror planes should be the same as the degree of the rotation axis. For instance, parallel to a 4-fold axis there should be 4 mirror planes ($4mm$); parallel to a 6-fold axis there should be 6 mirror planes ($6mm$); and so on (Fig. 1.15 a to d). If the mirror planes are perpendicular to the rotation axis (second case), then we get $2/m$, $4/m$ and $6/m$ for even axes and $1/m = m = \bar{2}$ and $3/m = \bar{6}$ for the odd axes. On the other hand, if there is an X-fold axis with X number of mirror planes parallel to it, and to this combination if a mirror plane is added perpendicular to the X-fold axis, it will generate X number of 2-fold axes

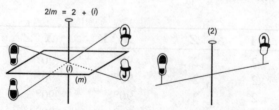

Fig. 1.14 Combination of an inversion axis ($\bar{1}$) or the centre of symmetry with a 2-fold axis generate a mirror plane ($2 + \bar{1} = 2/m$). Likewise $4 + \bar{1} = 4/m$ and $6 + \bar{1} = 6/m$ (not shown in the figure)

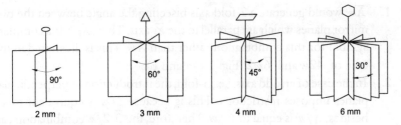

Fig. 1.15 Combination of mirror planes with proper rotation axis. Note that n-number of planes pass through a n-fold axis.

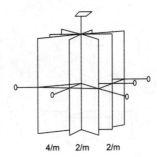

4/m   2/m   2/m

Fig. 1.16 Combination of mirror planes with a n-fold proper rotation axis. Note the generation of n-number of diagonals perpendicular to n-fold axis

perpendicular to the X-fold axis. Each of the parallel mirror planes contains one 2-fold axis. For example, 4*mm* on introduction of a perpendicular plane becomes $4/m\,2/m\,2/m$ or $4/mmm$ (Fig. 1.16). The above discussion leads to the general statement that a combination of any two of the three symmetry elements, namely a rotation axis of even degree, a mirror plane perpendicular to it and an inversion centre (centre of symmetry), leads to the third member of this trio being automatically generated.

1.6.6 Combination of an Improper Axis with a Proper Axis or Mirror Plane

Referring back to Table 1.1, one may replace two of the even-fold axis in each of the X, Y and Z combinations with their corresponding improper counterparts. This applies to the first five sets of Table 1.1. For example, 222 becomes $\overline{2}\overline{2}2$. Since $\overline{2}$ is less preferred than its equivalent symbol "m", $\overline{2}\overline{2}2$ becomes *mmm*. Likewise, 422 becomes $\overline{4}22$ or 4*mm*, depending upon whether the improper axis introduced was in the place of a 4-fold or a 2-fold axis.

When the improper axis is combined with mirror planes, a few special cases need attention.

1. $\overline{X}m$ would generate a 2-fold axis bisecting the angle between the two mirror planes which are parallel to the $\overline{X}$ axis. Thus, $\overline{X}m2$ is a complete symbol for this combination rather than Xm. This is illustrated in the case of $\overline{4}2m$ and $\overline{6}m2$ (Fig. 1.17a and b).

2.   In the case of an odd axis, i.e., 3-fold, the introduction of perpendicular planes imposes restrictions. This is because $3/m$ is equivalent to $\overline{6}$; besides, $\overline{3}/m$ is equal to $\overline{6}/m$. Therefore, the $\overline{3}\ 2/m$ combination can exist which has no mirror plane perpendicular to the $\overline{3}$ axes, as shown in Fig. 1.17c.

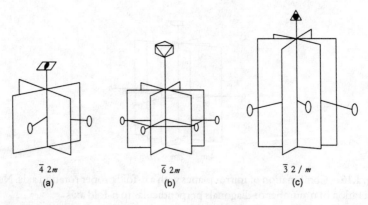

$\overline{4}\ 2m$ $\overline{6}\ 2m$ $\overline{3}\ 2/m$
(a)   (b)   (c)

Fig. 1.17   Combination of improper axis with mirror planes, i.e., $\overline{3}$, $\overline{4}$ and $\overline{6}$ axis with vertical planes. Note the generation of the 2-fold axis

3.   The combination of axes in classes 23 and 432 (Fig. 1.13) have to be understood with a guide cube, even when an inversion centre is added or a proper axis is being replaced by an improper axis. Thus, on addition of a centre of symmetry to 23 it becomes $2/m\overline{3}$ and 432 will become $4/m\overline{3}\ 2/m$. If the three 4-fold axes are replaced by three 2-fold axes, then all the 2-fold axes in the 432 combination will disappear to form $\overline{4}\ 3m$ (Figs. 1.18a–c). Besides, the centre of symmetry is also removed, leaving the only possible combination of $\overline{4}\ 3m$. In this combination it may be noted that there are three $\overline{4}$ axes, four 3-fold axes and six mirror planes.

    The introduction of a mirror plane perpendicular to an improper axis $(\overline{X}/m)$ would not generate any new combination. For example, $\overline{2}/m$ is same as m or $\overline{3}/m = 6/m$, $\overline{4}/m = 4/m$ and $\overline{6}/m = \overline{6} = 3/m$.

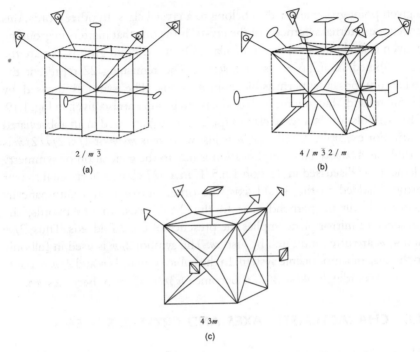

2 / m 3̄

(a)

4 / m 3̄ 2 / m

(b)

4̄ 3m

(c)

Fig. 1.18 Combination of symmetry elements in cubic classes and generation of additional symmetry elements from 23 and 432 classes. (a) Addition of a centre of symmetry to 23 combination generates three mirror planes with 2/m 3̄ symmetry. Combining the centre of symmetry to 432 combinations leads to the generation of nine mirror planes (three non-diagonal +6 diagonal), the final symmetry being 4/m 3̄ 2/m. (c) If 4̄ axis replace 4-fold axis in 432 combination, six diagonal planes are generated with the symmetry combination 4̄ 3m.

1.7 32 PERMISSIBLE POINT GROUPS

All the symmetry elements that a given crystal possesses will pass through a common point in the crystal. This common point does not move in space while performing symmetry operations of any kind. That means there is no translation involved for this common point of intersection of the symmetry elements. Therefore, the possible combinations of compatible symmetry elements discussed in the previous sections are called point groups. Figure 1.19 illustrates such possible combinations of symmetry elements, omitting the equivalent and, hence repetitive, combinations. One can find only 32 such non-identical combinations, known as the 32-point groups. In Crystallography, these point groups are synonymous with 32 crystal classes. This is because crystals possessing external symmetry elements identical to the symmetry of a given point group are said to belong to a crystal class. In other words, this collective external symmetry of the crystal is the one that matters in grouping a given crystal into an appropriate class. Of the 32 point groups, ten are the uncombined X and $\overline{X}$ axis themselves. The remaining 22 represent the maximum possible permissible combinations. The symbols devised by Hermann and Mauguin for each of the point group are also given in Fig. 1.19. The symbols for the crystal classes (point groups) are used in an abbreviated form. For example, $2/m\,2/m\,2/m$ is just written as mmm or $4/m\,2/m\,2/m$ is written as $4/mmm$. This simplification is due to the generation of symmetry elements as discussed in Section 1.6.5. If in a 422 class, a horizontal mirror plane is added to the 4-fold axis, the other mirror planes automatically generated will be perpendicular to the 2-fold axis. In otherwords, the presence of mirror plane implies the presence of the 2-fold axis. Thus, $2/m$ and $4/m$ are often indicated by just m. The symbol $2/m$ is used in full only in the case of a monoclinic system. In the cubic system, $4/m$ and $2/m$ become just m. Accordingly, $4/m\,\overline{3}\,2/m$ becomes $m3m$ or $2/m\,\overline{3}$ becomes $m3$.

1.8   CHARACTERISTIC AXES AND CRYSTAL SYSTEMS

All the 32 point groups may be arranged into 7 systems characterized by the presence of a specific symmetry axis, either single or in multiples. The symmetry axes so chosen are referred to as the characteristic axes. Table 1.2 gives the most widely chosen characteristic symmetry axes. Accordingly, the grouping of the 32 crystal classes leads to seven crystal systems whose names are also listed in Table 1.2. The table is introduced by the convention followed by the European Crystallographers. Their American counterparts, however, use only six crystal systems, because the rhombohedral (trigonal) system is considered as a sub-group of the hexagonal system. The reasoning is that the 3-fold axis is a sub-multiple of the 6-fold axis. The last column in Table 1.2 represents the maximum possible symmetry elements in a given system.

1.9   SCHOENFLIES NOTATIONS

In the earlier literature, other notations were used in Crystallography. Most of them became obsolete because of the versatility of the Hermann–Mauguin notations. However, the notations by Schoenflies still find use, mostly by spectroscopists and can be found in chemical literature. Therefore, it is imperative to give the equivalent notation of Schoenflies in addition to those

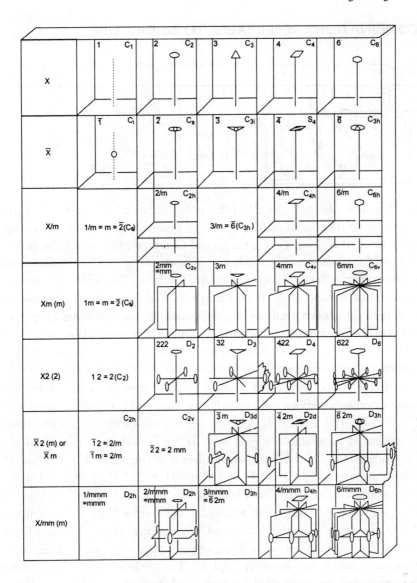

Fig. 1.19 The symmetry elements in 32 point groups derived by the permissible combinations of symmetry operations X (proper rotation axes), $\overline{X}$ (improper rotation axes) and m (mirror planes). The combinations of symmetry elements understood with a guide-cube belong to the cubic system, and are hence shown separately in the next page. The Hermann–Mauguin and Schoenflies notations are also shown. Boxes without figures indicate the repeating classes

COMBINATIONS REFERRABLE TO GUIDE-CUBE

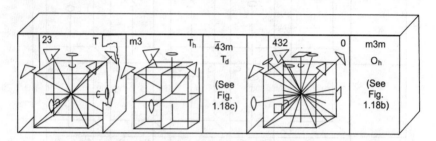

by Hermann–Mauguin. In the Schoenflies notation, C stands for cyclic symmetry of the axis of X-fold rotational symmetry, wherein the subscript is the value of X. Thus C_1, C_2, C_3, C_4 and C_6 indicate the crystal classes 1, 2, 3, 4 and 6 respectively. The symbol 'i' indicates the centre of symmetry and s stands for reflection (Spiegelung in German) Therefore, C_i and C_s means $\overline{1}$ and m respectively. If a cyclic axis (X-fold) has a mirror plane normal to it, then the general symbol is C_{nh}. For example, C_{2h} represents $2/m$, C_{3h} is $3/m$ or $\overline{6}$. The symbol h indicates a horizontal mirror plane, whereas v indicates the vertical mirror plane. Schoenflies used the concept of rotoreflection (rotation combined with reflection) to explain the improper axis. Accordingly, the current symbol of $\widetilde{1}$, $\widetilde{2}$, $\widetilde{3}$, $\widetilde{4}$ and $\widetilde{6}$ (~ to be pronounced as 'tilde'), the corresponding notations by Schoenflies are S_1, S_2, S_3, S_4 and S_6. Here, $S_1 = C_s$, $S_2 = C_i$, $S_3 = C_{3h}$ and $S_6 = C_{3i}$. S_4 is a new class with Schoenflies whose equivalent in Hermann–Mauguin is $\overline{4}$. If an X-fold axis has a 2-fold axis perpendicular to it, Schoenflies symbolizes it as Dn, where $n = X$. Thus, D_2 is equivalent to 222, $D_3 = 32$, $D_4 = 422$ and $D_6 = 622$. If a horizontal mirror plane is added to a D_n class, they become D_{nh}. Thus, $D_{2h} = 2/m\,2/m\,2/m$, $D_{3h} = \overline{6}2m$, $D_{4h} = 4/m\,2/m\,2/m$, and so on. If a mirror plane is diagonal to a 2-fold axis then the symbol 'd' is used. For example, $D_{2d} = \overline{4}2m$ and $D_{3d} = \overline{3}2/m$. The Schoenflies equivalence of the 32 point groups for Hermann–Mauguin notations are shown in Table 1.2. The symbols T and O which stand for tetrahedral and octahedral symmetry, belong to the cubic system. Thus, $O = 432$ and $T = 23$.

Table 1.2 The 32 point groups classified under seven crystal systems

Crystal System	Characteristic Symmetry Elements	No. of Crystal Classes	Axes of Symmetry (f =fold)				Planes of Symmetry	Centre of Symmetry	International Class Symbol		Total Symmetry Elements
			2-f	3-f	4-f	6-f			H.M	S.F	
Triclinic	1 or $\bar{1}$	2	–	–	–	–	–	–	1	C1	1
			–	–	–	–	–	Present	$\bar{1}$	Ci	1
Monoclinic	$2, 2/m$ or m	3	1	–	–	–	–	–	2	C_2	1
			–	–	–	–	One plane ⊥ to *b*-axis	Absent	m	C_s	1
			1	–	–	–	One plane ⊥ to 2-f axis	Present	$2/m$	C_{2h}	2
Ortho-rhombic	Three 2-fold or mm2	3	3	–	–	–	–	Absent	222	D_2	3
			1	–	–	–	Two Vertical planes parallel to 2-f axis	Absent	mm (mm2)	C_{2v}	3
			3	–	–	–	Three planes ⊥ to three 2-f axes	Present	mmm	D_{2h}	6
Hexagonal	6-fold or $\bar{6}$-axis	7	–	–	–	1	–	Absent	6	C_6	1
			6	–	–	1	–	Absent	62	D_6	7
			–	–	–	1	One plane ⊥ to 6-f axis	Present	$6/m$	C_{6h}	3
			–	1	–	–	One plane ⊥ to 3-f axis	Absent	$\bar{6}\left(=\dfrac{3}{m}\right)$	C_{3h}	2

(continues....)

(...contd)

Table 1.2 The 32 point groups classified under seven crystal systems

Crystal System	No. of Crystal Classes	Characteristic Symmetry Elements	Axes of Symmetry (f = fold)				Planes of Symmetry	Centre of Symmetry	International Class Symbol		Total Symmetry Elements
			2-f	3-f	4-f	6-f			H.M	S.F	
			–	–	–	1	Six vertical planes parallel to 6-f axis	Absent	$6mm$	C_{6v}	7
			3	1	–	–	One plane ⊥ to 3-f axes and three planes parallel to 3-f axes	Absent	$\bar{6}2m$	D_{3h}	8
			6	–	–	1	One plane ⊥ to 6-f axis and six planes ⊥ to 2-f axes	Present	$6/mmm$	D_{6h}	15
Trigonal	5	3-fold or $\bar{3}$-axis	–	1	–	–	–	Absent	3	C_3	1
			–	1	–	–	–	Present	$\bar{3}$	C_{3i}	2
			3	1	–	–	–	Absent	32	D_3	4
			–	1	–	–	Three planes parallel to 3-f axis	Absent	$3m$	C_{3v}	4
			3	1	–	–	Three diagonal planes parallel To 3-f axis and	Present	$\bar{3}m$	D_{3d}	7

(continues...)

(...contd)

Table 1.2 The 32 point groups classified under seven crystal systems

Crystal System	Characteristic Symmetry Elements	No. of Crystal Classes	Axes of Symmetry (f =fold)				Planes of Symmetry	Centre of Symmetry	International Class Symbol		Total Symmetry Elements
			2-f	3-f	4-f	6-f			H.M	S.F	
							⊥ to the 2-f axis				
Tetragonal	Single 4-fold or $\bar{4}$-axis	7	–	–	1	–	–	Absent	4	C_4	1
			–	–	1	–	–	Absent	$\bar{4}$	S_4	1
			–	–	1	–	One plane ⊥ to 4-f axis	Present	4/m	C_{4h}	3
			–	–	1	–	Four planes parallel to 4-f axis	Absent	4mm	C_{4v}	5
			2	–	1	–	Two diagonal planes parallel to $\bar{4}$-f axis	Absent	$\bar{4}2m$	D_2d	5
			4	–	1	–	–	Absent	422	D_4	5
			4	–	1	–	Five planes ⊥ to each of the rotational axis	Present	4/mmm	D_{4h}	11
Cubic	Four 3-fold or $\bar{3}$-axis	5	3	4	–	–	–	Absent	23	T	7
			3	4	–	–	Three planes ⊥ to 2-f axis	Present	m3 (2/m $\bar{3}$)	T_h	9

(continues...)

(...contd)

Table 1.2   The 32 point groups classified under seven crystal systems

Crystal System	Characteristic Symmetry Elements	No. of Crystal Classes	Axes of Symmetry (f =fold)				Planes of Symmetry	Centre of Symmetry	International Class Symbol		Total Symmetry Elements
			2-f	3-f	4-f	6-f			H.M	S.F	
			3	4	–	–	Six diagonal planes ⊥ to 2-f axes	Absent	$\bar{4}3m$	T_d	14
			6	4	3	–	–	Absent	432	O	13
			6	4	3	–	Nine planes 3 ⊥ to 4-f axis and Six planes ⊥ to 2-f axes	Present	$m3m$ $(4/m\,\bar{3}\,2/m)$	O_h	23

Description of Crystals

2.1 INTRODUCTION

Naturally occuring crystals with well-bound geometrical faces are rather rare. They may or may not possess smooth faces and are accordingly called euhedral or anhedral crystals. If they are partly bounded by smooth faces, they are called subhedral crystals. Even in euhedral crystals there may be ill-shaped and distorted crystals which makes it difficult to recognize all the possible symmetry elements and identify the crystal class. The distortion in crystals essentially arises because of improper growth conditions such as the non-availability of constituents to all the faces equally, or the presence of impurities or the external physical constraints prevailing during the growth. In the absence of good euhedral crystals of discernible symmetry, the relative ratios of length, breadth and height of a crystal, the relative areas of the faces, the constancy of the angular relations between the faces (interfacial angles), and so on, become significant. This point will be further developed in a later section. In spite of these difficulties of quantifying the external geometrical features in natural crystals, certain generalizations have been drawn taking into consideration the ideal polyhedra. For this purpose, the concepts of solid geometry have been used in Crystallography. The idea of reference crystallographic axes and their interaxial angles have been found convenient in describing the crystals. The polyhedra used in these considerations are of conceptual importance in geometrical (morphological) Crystallography. However, these concepts found justification after the advent of X-ray diffraction and its application to crystals, wherein the *unit cell* as a building block has been amply substantiated (Chapter 4). The ideal polyhedra discussed in the following sections is rarely found in nature.

2.2 CRYSTALLOGRAPHIC AXES

A set of reference axes chosen for convenience are called the crystallographic axes. Conventionally, these axes should meet the following requirements:

a) There are normally three crystallographic axes (4 in a hexagonal system).
b) If possible the axes should be mutually perpendicular.
c) As far as possible they should coincide with the symmetry axes.
d) In the absence of an axis of symmetry, they coincide with the normal to the symmetry planes.
e) In the absence of both axes and planes of symmetry, the intersection-edges between faces of the largest area are considered to choose the direction of the axes.

Incidentally, the seven crystal systems described in terms of the characteristic symmetry of axes, also show an individual characteristic set of crystallographic axes and interaxial angles. The following brief treatment is in terms of the crystallographic axes of the seven crystal systems.

Triclinic System: This is so named, because the three crystallographic axes are inclined to one another at angles other than 90°. The crystal classes under this system $(1, \bar{1})$ lack symmetry axes and symmetry planes. Conventionally, the three crystallographic axes are designated as a, b and c. These respectively correspond to: i) back to front (away and towards the observer), ii) side to side (right to left), and iii) vertical (up and down). These axes also have +ve or −ve signs ascribed to them. Towards the front, to the right and to the top are designated as $+a$, $+b$ and $+c$ respectively, while the opposite directions bear the corresponding −ve sign. The interaxial angles are also designated by convention. The angle between $+b$ and $+c$ axes is called γ, that between $+c$ and $+a$ as β and $+a$ and $+b$ as α. Fig. 2.1a shows the axial system in relation to a triclinic polyhedron. The above scheme of designating the crystal axes and interaxial angle holds good for all other systems (except the hexagonal).

Monoclinic System: The crystal classes falling under this system are: $2/m$, 2 and m. Here, the angle β is not equal to 90°, whereas α and γ are equal to 90°. Therefore, the a- and c-axis are inclined to each other. By convention, c is held vertical such that the $+a$ is inclined to the observer. The 2-fold axis is the b axis while the mirror plane, if present, is normal to the 2-fold axis on the a-c plane (Fig. 2.1b).

Orthorhombic System: The three crystallographic axis are mutually perpendicular and the angles $\alpha = \beta = \gamma = 90°$. The $2mm$ (mm), 222 and mmm are the crystal classes belonging to this system. In the first case, the 2-fold axis can coincide with the a-axis ($2mm$), the b-axis ($m2m$) or the c-axis ($mm2$). The 2-fold axis is conventionally chosen to coincide with the crystallographic axes (Fig. 2.1c).

Tetragonal System: The difference between this and the orthorhombic system is that the c-axes are chosen to coincide with the 4-fold axis. The remaining two crystallographic axes are mutually perpendicular, interchangeable and lie in a plane perpendicular to the *c*-axis (Fig. 2.1d). In the class $\overline{4}2m$, the a_1 and a_2 axes do not coincide with the 2-fold axes present. In the case of 4 *mm* and 4/*mmm*, the a_1 and a_2 crystallographic axes coincide with any one of the mirror planes parallel to the *c*-axis. The crystal classes belonging to these systems are 4, $\overline{4}$, 4/*m*, 4*mm*, $\overline{4}2m$, 42 (or 422) and 4/*mmm*.

Cubic System: In the cubic system, the *a*, *b* and *c* are chosen to coincide with the three 4-fold axes. In the absence of 4-fold axes, as in the case of the crystal class 23, the crystallographic axes are chosen to coincide with the 2-fold or $\overline{4}$ -fold axes. The three crystallographic axes are not only mutually perpendicular but are the same in linear dimensions (designated a_1, a_2 and a_3). These axes are called isometric, meaning "equal measure". The crystal classes belonging to the isometric system are 23, $\overline{4}$ 3*m*, *m*3 (= 2/*m*$\overline{3}$), 43 (or 432) and *m*3*m* (= 4/ *m* $\overline{3}$ 2/*m*). As mentioned in section 1.8, the characteristic symmetry axes are four 3-fold axes, although they are never chosen to coincide with any of the crystallographic axis (Fig. 2.1e).

Hexagonal System: An ideal hexagonal prism needs four axes for its full description. Of these, the *c*-axis coincides with the 6 or $\overline{6}$ axis while the other three interchangeable axes are designated as a_1, a_2 and a_3, which are perpen-

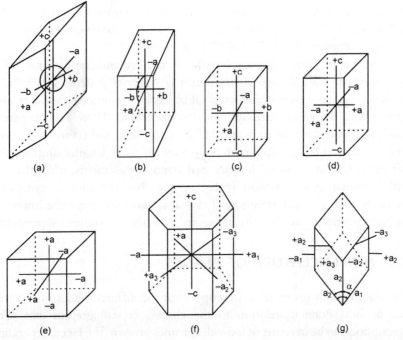

Fig. 2.1 The axial notations in the seven crystal systems: (a) Triclinic (b) Monoclinic (c) Orthorhombic (d) Tetragonal (e) Cubic (f) Hexagonal (g) Trigonal.

perpendicular to the *c*-axis and are 60° apart in the same (horizontal) plane. A conventional hexahedron of a_1, a_2 and a_3 is in accordance with that shown in Fig. 2.1f. Again, by convention these axes bisect the angle between the two adjacent 2-fold axes, if present. The crystal classes belonging to the hexagonal system are 6, $\overline{6}$, 6/*m* 6*mm*, $\overline{6}$ 2*m*, 62 (or 622) and 6/*mmm*.

Trigonal System: As mentioned in Section 1.8, the characteristic symmetry axis of a trigonal system is 3 or $\overline{3}$. The choice for the crystallographic axes with respect to the symmetry element are two. According to the American crystallographers, the 3-fold axis coincides with the *c*-axis and the three a_1, a_2 and a_3 axes correspondingly coincide with the 2-fold axis in the horizontal plane (Fig. 2.1g). The choice of axes followed in this figure makes it clear that the trigonal system should be a part of the hexagonal, except that the triad axis in the former is replaced by the hexad axis in the latter. Therefore, these are two sub-systems (namely trigonal and hexagonal) in the larger hexagonal system. However, the European crystallographers preferred to keep the trigonal as a separate system and have chosen another set of crystallographic axis. Accordingly, the a_1, a_2 and a_3 axes coincide with the three edges (Fig. 2.1g) of the rhombohedra. In this scheme, the three axes are interchangeable but do not coincide with any symmetry axes. The interaxial angle between the chosen 3 axes is the same (designated as α). The value of α is other than 90° and, hence, it differs from the cubic system. The crystal classes belonging to the trigonal system are 3, $\overline{3}$, 3*m*, 32, and $\overline{3}$ *m* (or $\overline{3}$ 2*m*).

In summary, these selective lengths of the crystallographic axes (axial ratios) and the *interaxial angles* define the different crystal systems. Incidentally, the seven crystal systems, characterized by distinct symmetry elements, also concur with the seven axial systems described here. There is an apparent relationship between the axial system (crystal system) and crystal symmetry. The system with the maximum variations in axial lengths and angular variations (triclinic system) has minimal symmetry elements, while the one with minimal axial variation (cubic system) has the highest symmetry elements. The above description of crystal systems was from the lowest to highest symmetry (treating the hexagonal and trigonal systems separately).

2.3   NOMENCLATURE OF PLANES

An unambiguous geometrical description of the different faces of a crystal can be best done in relation to the chosen crystallographic axis. The description can be in terms of individual names given to the faces, depending on their disposition with reference to the given crystallographic axis. Alternatively, it can be made more precise in terms of the relative lengths of the intercepts the face makes with the crystallographic axes. The latter method became widely accepted and has therefore been developed first, followed by the descriptive nomenclature.

2.3.1 Weiss Parameters

These are the relative numbers of units at which a given face cuts the crystallographic axes. In order to determine the Weiss parameter of any face, a unit face has to be chosen on the crystal which has the largest area, is inclined to all the crystallographic axes and cuts them at unit distances. This means that the Weiss parameters of a unit face will be $1a : 1b : 1c$. Consider any other face with a different inclination and hence differing in the lengths of intercepts. These lengths are expressed as ratios with respect to the intercept made by the unit face. Thus, the Weiss parameters $3a : 3b : 1c$ for a face in an orthorhombic crystal means that this face intercepts the a- and b-axis at thrice the value of the unit lengths a- and b-, and cuts the c-axis at the same unit distance. If the face is parallel to a given axis, it is assumed to intercept this axis at infinity and the corresponding symbol carries '∞' sign. To illustrate this point, the following example may be considered. Figure 2.2 shows a crystal of sulphur where the larger face m is chosen as the unit face; then the lengths of intercept that this face makes with the a-, b- and c-axis can be taken as the unit lengths, in terms of which the intercepts of other faces are defined in Weiss parameters. If the intercepts are assumed to be 7 : 8 : 16 cm for face m,

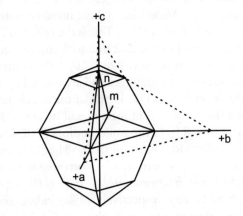

Fig. 2.2 The nomenclature of faces (planes) using Weiss parameters. The face m is taken as a unit face and its intercept length with the crystallographic axis as unit distances. The Weiss parameters of face n is with reference to the unit face (See text).

and $14 : 18 : 12$ cm for the face *n*, then the Weiss parameter for the face *n* are $14/7 : 18/8 : 12/16 = 2a : 2.25b : 0.75c$.

2.3.2 Miller Indices

Miller indices are simply the reciprocals of Weiss parameters except that the *a*, *b* and *c* denoting the axes are omitted. Besides, the ratios are cleared of the

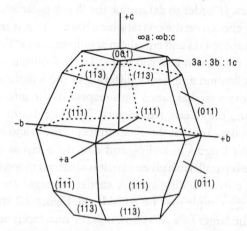

Fig. 2.3   The Miller indices of faces derived from the Weiss parameter.

fraction by multiplying them by a suitable number to yield whole number values. For example, if the Weiss parameters for a given face are $3a : 3b : 1c$ then the reciprocals of these parameters are $1/3$, $1/3$ and 1 on being multiplied by 3 yield 1, 1, 3. Miller placed these numbers without the commas and within parenthesis to obtain (113). This is the called Miller Index, usually read as 'one one three' and not as one hundred and thirteen. If any one of these indices has a two digit number like 1, 10, 2, the commas are still used. Miller indices also do not have multiples and are cleared by common factors. For example, (336) is written as 112. If one of the axes is not intercepted by a given plane then the reciprocal of infinity will be zero. Hence, $2a : 2b : \infty$ c in the Weiss symbol will have the Miller indices $(1/2, 1/2, 1/\infty)$ which is equal to (110). The Miller index of a unit face will be (111). In general, a Miller index is referred as (*hkl*) where *h*, *k*, *l* correspond to the reciprocated intercepts cleared of the fractions. (0*kl*), (*h0l*) (*hk*0) represent the faces parallel to the *a*-, *b*- and *c*-axes, respectively. Miller indices such as (*hhl*) mean that the *h* and *k* have the same values. In this sense, (*hhh*) means that the face intercepts *a*, *b* and *c* are at the same distance. When the plane makes the intercept on the negative segments of the crystallographic axis, the corresponding Miller indices will carry a negative sign which is placed on the top of the corresponding symbol by a short bar. For example, $11\overline{3}$ represents a face which intercepts the c-axis on the negative segment. It reads "one, one, bar three" (Fig. 2.3).

2.3.3 Miller–Bravais Indices

As mentioned in Section 2.2, the crystals of the hexagonal system have four crystallographic axes. Therefore, the indices to various crystal faces involve four intercepts. Such a set of four crystallographic axes was originally developed by Bravais, the corresponding Miller indices adopted for the hexagonal system are called Miller–Bravais indices. The four axes involved are a_1, a_2, a_3 and c. The Weiss parameter of a unit face will be $1a_1 : 1a_2 : 2a_3 : 1c$, which means it cuts a_1-, a_2- and c-axes in positive directions at unit distances and cuts the a_3-axis at twice the distance at the negative end. Any plane cutting the $+a_1$ and $+a_2$ can only intercept on an $-a_3$-axis. Alternatively, a plane intercepting the $-a_1$ and $-a_2$-axes can only intercept on $+a_3$-axis. In general, for any Miller–Bravais index, the sum of first three indices will be equal to zero. Thus $(11\overline{2}1)$ refers to a face that cuts the positive ends of a_1 and a_2 at unit distances, at half the distance at negative $-a_3$ and unit distance at $+c$-axis. In general, the Miller–Bravais Index is referred to $hk\overline{i}l$, where $h + k + i$ should be equal to zero. Some crystallographers also use this indices as $h, k, *l$ because of the equality of the i value once the h and k values are known. Some even eliminate this * sign.

2.3.4 General Names of Crystal Faces or Forms

A crystal form is an assemblage of crystal faces which are related to one another through a symmetry operation. Hence, a crystal form is related to the crystal class or point group to which it belongs. The following nomenclature has been in use in crystallography for broadly grouping the crystal forms. If a crystal lacks any symmetry element (as in class 1) each of the single faces is called a pedion (Fig. 2.4a). If the plane of the face is related to an inversion symmetry (as in class $\overline{1}$), the form is called a pinacoid (Fig. 2.4b). If a pair of faces is related to a two-fold axis and if these faces are not parallel, they are called the sphenoids (Fig. 2.4c). The pair of non-parallel faces which are related through a mirror are called domes (Fig. 2.4d). The dome faces differ from the sphenoids, in that the former contain the pair of faces with right-hand to left-hand relations. Consider that there are two pairs of planes (a_1, a_3, a_2, a_4) interrelated to one another by a 2-fold axis and at the same time through a mirror plane perpendicular to the 2-fold axis ($2/m$ class), then the faces are called the prism faces (Fig. 2.4e). Prism faces will be parallel to any one of the crystallographic axes (Fig. 2.5). Three or more inclined faces which intersect on a common point on any one of the crystallographic axes are called the pyramids. They are inclined to all the crystallographic axes (Fig. 2.6). The common forms shown in Figs. 2.5 and 2.6 all belong to non-isometric systems (See Table 2.1). The forms of isometric classes are different because

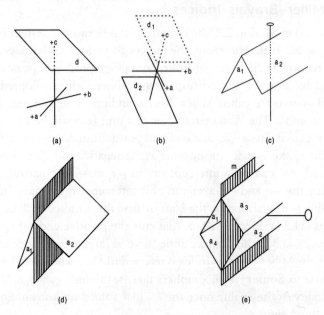

Fig. 2.4 The general forms of faces: (a) Pedion (single face) (b) Pinacoid (open form) (c) Sphenoid (dihedron; open form) (d) Dome (open form) (e) Prism (open form).

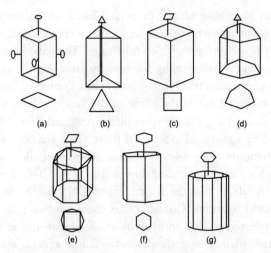

Fig. 2.5 The common forms of non-cubic crystals: (a) Rhombic prism (b) Trigonal prism (c) Tetragonal prism (d) Ditrigonal prism (e) Ditetragonal prism (f) Hexagonal prism (g) Dihexagonal prism.

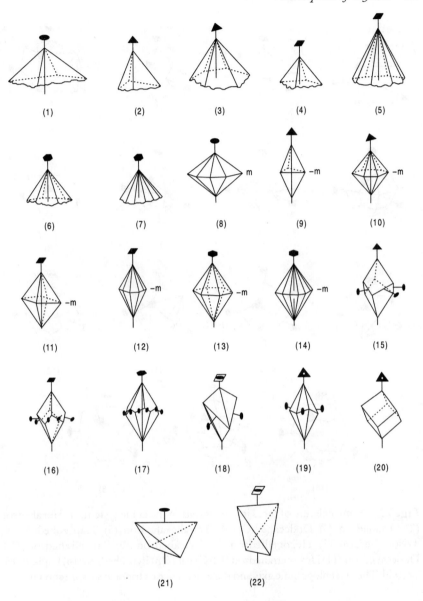

Fig. 2.6 Non-parallel faces in non-cubic crystals: (1) Rhombic pyramid (2) Trigonal pyramid (3) Ditrigonal pyramid (4) Tetragonal dipyramid (5) Ditetragonal pyramid (6) Hexagonal pyramid (7) Dihexagonal pyramid (8) Rhombic dipyramid (9) Trigonal dipyramid (10) Ditrigonal dipyramid (11) Tetragonal dipyramid (12) Ditetragonal dipyramid (13) Hexagonal dipyramid (14) Dihexagonal dipyramid (15) Trigonal trapezohedron (16) Tetragonal trapezohedron (17) Hexagonal trapezohedron (18) Tetragonal scalenohedron (19) Hexagonal scalenohedron (20) Rhombohedron (21) Rhombic disphenoid (22) Tetragonal disphenoid.

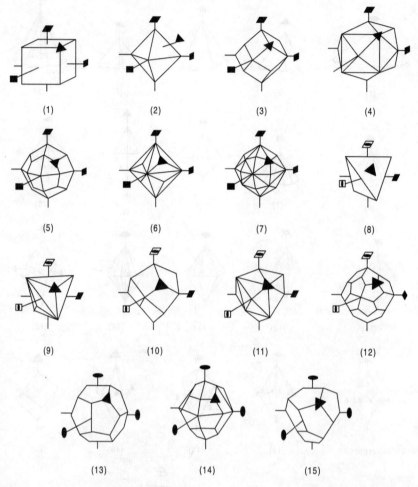

Fig. 2.7 Nomenclature of faces in the crystals of the cubic system: (1) Hexahedron (2) Octahedron (3) Dodecahedron (4) Tetrahexahedron (5) Trapezohedron (6) Trisoctahedron (7) Hexoctahedron (8) Tetrahedron (9) Tristetrahedron (10) Dodecahedron (11) Hexa-tetrahedron (12) Gyroid (13) Pyritohedron (14) Diploid (15) Tetroid. The crystallographic axis and the emergence of trigonal axis are shown

they possess four 3 or $\bar{3}$ axes and these forms are shown in Fig. 2.7. (See also Table 2.2)

2.3.5 Forms in different Crystal Systems

As mentioned in Section 2.3.4, a crystal form is an assemblage of faces which are symmetry related. The aim of this section is to familiarize the different euhedral crystal forms belonging to the 32 crystal classes. This is more important for students of mineralogy rather than to the readers from other disciplines. Conventionally, the crystal forms are grouped into two categories: (a) special forms (b) general forms. If the crystal faces are parallel to one or more of the symmetry elements, they are said to be special forms. The faces with indices $(h00)$, $(hk0)$, etc., are of this group. The faces which are not parallel to or perpendicular to any of the symmetry elements are called the general forms. The Miller indices of such faces will be (hkl) or $(h\bar{k}l)$. The form symbols are represented by curved brackets { }. For example, {111} means a set of planes of indices (111), $(\bar{1}11)$, $(1\bar{1}1)$, $(11\bar{1})$, $(\bar{1}\,\bar{1}1)$, etc. In a given crystal system, the class with the highest number of permissible symmetry elements is said to be *holohedral*. The holohedral class will possess the maximum number of forms (both general and special). As the number of permissible symmetry elements decreases, the number of faces belonging to each form also decreases as in the case of *hemihedral* and *tetartohedral* forms. The reduction in the so-called hemihedral form is by half the number of holohedral forms and in tetartohedral forms is by 1/4 the numbers of holohedral forms.

2.4   CRYSTAL PROJECTIONS

The disposition of the faces on crystals in relation to symmetry as well as crystal axes cannot be quantitatively represented on a paper, however skilful the perspective drawings and plans are. The angular relationships of faces are very critical in the study of crystals. This difficulty of representations was overcome by projecting the 3-dimensional relations onto a 2-dimensional plane. Such projections should not only carry information regarding the symmetry of crystals, but also vividly represent the angular relationships between the faces and axes. Three types of projections have been found to be very useful: a) spherical projections b) stereographic projections c) gnomonic projections. The stereographic and gnomonic projections are the derivatives of the spherical projection.

2.4.1 Spherical Projection

In a *spherical projection* (Fig. 2.8), the crystal is imagined to be placed in a sphere with the centres coinciding. The principal axis (c-axis) of the crystal usually coincides with the north–south diameter of the sphere. The normals to each of the faces are drawn in the upper hemisphere and extended until they intersect the surface of the sphere. The points of intersection are called the poles of the faces. Their distribution on the surface of the sphere represents the true disposition of planes on the crystal. The poles of those faces, which are parallel to themselves, fall on a great circle. A great circle is a trace of a plane on the surface of the sphere passing through the centre of the sphere and it represents a plane (possible symmetry plane) passing through the crystal. A *great circle* can be vertical, horizontal or inclined. If we slice the sphere along the plane which does not pass through the centre, the traces of such planes on the surface of the sphere are called the *small circles*. The intersection of the great circles is a possible face.

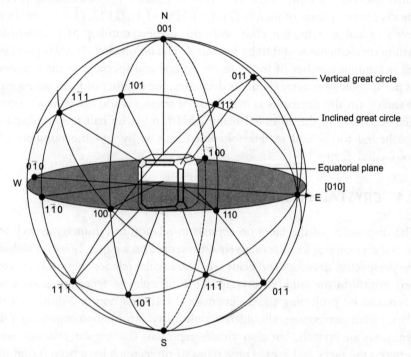

Fig. 2.8   Spherical projection. The representation of crystal faces on the surface of a sphere, as poles of the respective faces.

2.4.2 Stereographic Projection

The *equatorial plane* in Fig. 2.8 can be taken as the plane of projection. Instead of projecting the faces and planes on the surface of the sphere as poles and great circles respectively, they can be transferred onto the equatorial plane. This can be most faithfully done by joining every point on the northern hemisphere to the south pole (Fig. 2.9), such that the points of intersection on the equatorial plane now represent the transferred poles. The vertical great circles appear as diameters of the equatorial plane and the inclined great circles appear as arcs. The equatorial plane is then turned towards the observer to yield what is known as the stereographic projection (Fig. 2.10). In Fig. 2.10, we find that the basal plane (001) will be at the centre. This is true only in the case of orthogonal systems where the emergence of the *c*-axis coincides with the normal to (001) face. In a monoclinic system it slides down on the N-S diameter by an angle proportional to the angle β. In a triclinic system it even drifts from the N–S diameter. The poles of the faces parallel to the *c*-axis will fall on the perimeter of the horizontal great circle, also known as the plane of projection or the primitive circle in stereographic projection. All the poles of faces parallel to the *b*-axis lie on the N–S diameter (true only in the orthogonal and monoclinic systems), and those parallel to the *a*-axis will be on the E–W diameter (true only in the orthogonal system). The poles which fall within any of the quadrants will have Miller Indices (*hkl*), the sign and value of *h*, *k* and *l* depending upon the chosen quadrant. The disposition of the poles of faces on the stereographic projection plane in the different

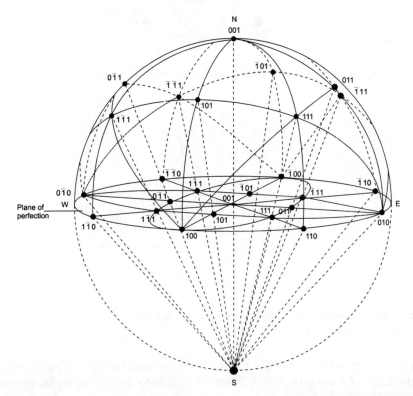

Fig. 2.9   Derivation of a stereographic projection from the spherical projection, where all the poles and their zones (great circles) from the northern hemisphere are transferred to the equatorial planes.

crystal systems will be clear in later sections, as it is related to the crystallographic axial angles. Whatever the distribution of poles, the relationship between the angle formed by two poles on the spherical projection and the correspondingly transferred poles on the stereographic projection is given by

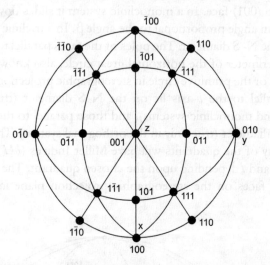

Fig. 2.10 Stereographic projection of cubic crystal with poles of cubic and dodecahedral faces.

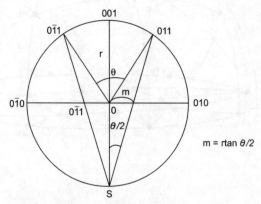

Fig. 2.11 The angular relation between two poles ($01\bar{1}$ and 011) in spherical and stereographic projection.

the relation: $m = r\tan\theta/2$ (Fig. 2.11). Thus, any angular relationship between the faces of a crystal is faithfully transferred onto the stereographic projection. The angle between the centre of projection to the point on the horizontal great circle is 90°. Any point which is 90° away from any great circle is called the pole of that great circle.

2.4.3   Plotting of Faces on Stereographic Projection

A Wulff's net is used for this purpose which is a graphical method for constructing the stereogram of a crystal. The net shown in Fig. 2.12 is a projection of the latitudes and longitudes of a sphere on its equatorial plane. If one views the vertical great circle of northern hemisphere in Fig. 2.9 from the south pole, it appears as a diameter on the equatorial plane. At the same time, the inclined great circles appear as arcs. Wulff's net is constructed with the projection of small and great circles. The interval between each is kept at

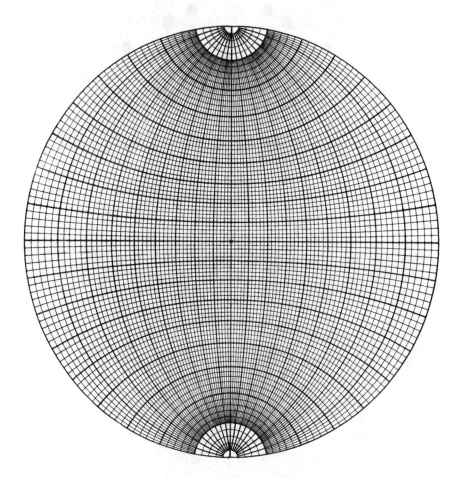

Fig. 2.12   Wulff's stereographic net. 1 division = 2°. The latitudes represent the vertical great circle at an interval of 2° where the longitudes are small circles.

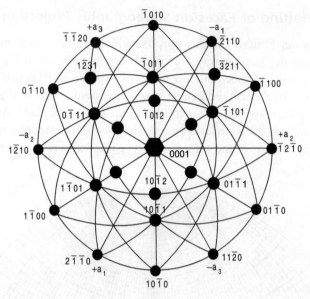

Fig. 2.13   Stereographic projection of a hexagonal crystal showing the possible poles of faces and their Miller indices.

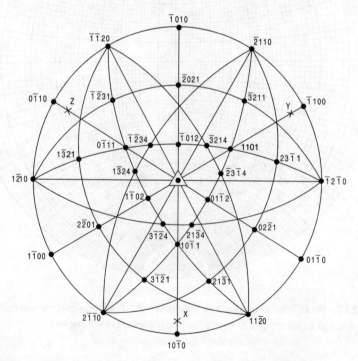

Fig. 2.14   Typical stereographic projection of a trigonal crystal of calcite.

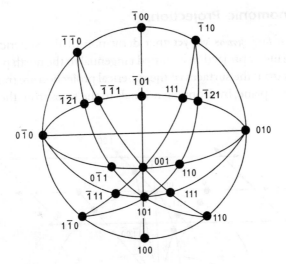

Fig. 2.15 A typical stereographic projection of a monoclinic crystal. Note the shift in the (001) pole from the centre.

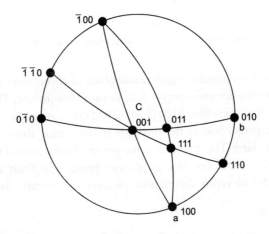

Fig. 2.16 A part of the stereographic projection of a triclinic crystal. Note the shift in the 100 and 001 poles.

5° or 2° depending upon the accuracy needed. Making use of the Wulff stereonet, the symmetry projection of a crystal as well as the projection of its faces can be represented without any distortion of the angular relationship. Figures 2.13 to 2.16 represent the stereographic projections of typical hexagonal, trigonal, monoclinic and triclinic crystals.

2.4.4   Gnomonic Projections

The *gnomonic projection* is yet another derivation of the spherical projection. Here, the plane of projection is shifted tangential to the north pole. The poles of the faces from the surfaces of the spherical projection are transferred onto this tangential plane, by dropping normals to the faces from the centre of the

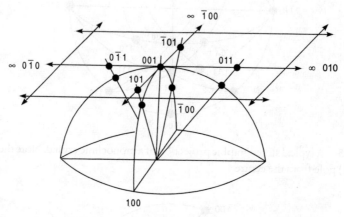

Fig. 2.17   Projection of poles on a tangential plane of gnomonic projection.

sphere until they intersect the gnomonic plane of projection (Fig. 2.17). The faces parallel to the common axis will all lie on a straight line. The faces which are parallel to the *c*-axis like (010) (110) (210), etc., cannot be represented on the gnomonic projection as the normals transforming them run parallel to the projection plane. This projection has several limitations of representation of all the faces. However, the gnomonic projection finds importance in interpreting the electron diffraction pictures of crystals as described in Chapter 4.

2.5   SYMMETRY PROJECTIONS OF 32 POINT GROUPS

The stereographic projections can also be made use of to represent the symmetry elements present in each of the 32 point groups. The minimum required face-poles are marked in order to account for every rotation axis and mirror plane present. In the figures that follow, the full circle and bold lines indicate the presence of mirror planes. The absence of a plane is indicated by a dotted circle or line. The face-poles corresponding to the upper hemisphere are marked with a cross while those on the lower hemisphere are shown as circles. The crystal, obviously, is oriented in such a way that its *c*-axis

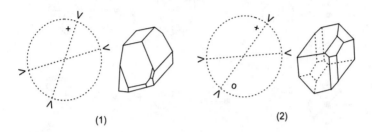

(1) (2)

Fig. 2.18 Crystal classes of the triclinic system: (1) Asymmetric class [1] sodium thiosulphate, each face is unique (2) Triclinic holosymmetric class [$\bar{1}$] axinite

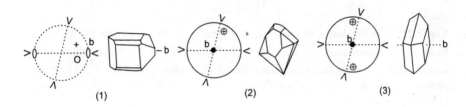

(1) (2) (3)

Fig. 2.19 Crystal classes of the monoclinic system:—the 2-fold axis coincides with the *b*-axis: (1) Sphenoidal class [2] tartaric acid (2) Clinohedral [*m*] $Na_2SiO_3 5H_2O$ (3) Holosymmetric [2/*m*] $CaSO_4 2H_2O$

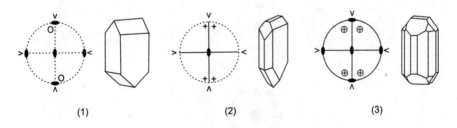

(1) (2) (3)

Fig. 2.20 Crystal classes of the orthorhombic system: (1) sphenoidal [222] $MgSO_4 7H_2O$ (2) Hemimorphic [2*mm* or *mm*] hemimorphite $Zn(OH)_2SiO_2$ (3) Holosymmetric [2/*mmm* or *mmm*] $PbSO_4$

is perpendicular to the plane of the paper and consequently the emergence of the *c*-axis coincides with the centre of the stereogram. The symmetric projections of the 32 point groups as they appear with increasing symmetry

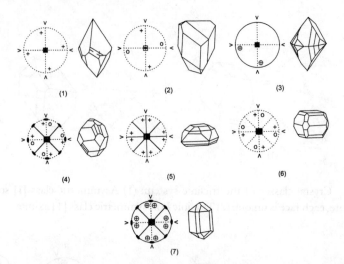

Fig. 2.21 Crystal classes of the tetragonal system: (1) Hemimorphic [4] Wulfenite $PbMoO_4$ (2) Sphenodial [$\bar{4}$] Cahnite $Ca_4B_2O_6(OH)_2 5H_2O$ (3) Bipyramidal [4/m] scheelite $CaWO_4$ (4) Bisphenoidal or scalenohedral - [$\bar{4}\ 2m$] chalcopyrite $CuFeS_2$ (5) Hemimorphite [4mm] diabolite. $2Pb(OH)_2 CuCl_2$ (6) Trapezohedral [42] $NiSO_4 6H_2O$ (7) Holosymmetric [4/mmm] cassiterite, SnO_2

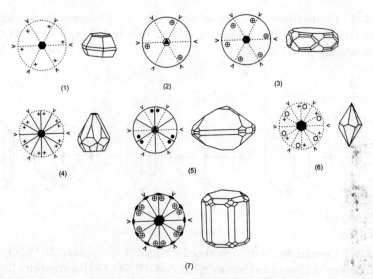

Fig. 2.22 Crystal classes of the hexagonal system—6 or $\bar{6}$ is the vertical c axis: (1) Hemimorphic [6] nepheline, $NaAlSiO_4$ (2) Trigonal bipyramidal [3/m = $\bar{6}$] lead germanate, $Pb_5Ge_3O_{11}$ (3) Hexagonal pyramidal [6/m] apatite ($CaF_2 Ca_4 (PO_4)_3$ (4) Dihexagonal hemimorphic [6mm] zincite, ZnO (5) Trigonal bipyramidal [62m] benitoite $BaTiSi_3O_9$ (6) hexagonal trapezohedral [622] β-quartz, SiO_2 (7) Holosymmetric [6/mmm] beryl, $Be_3Al_2Si_6O_{18}$

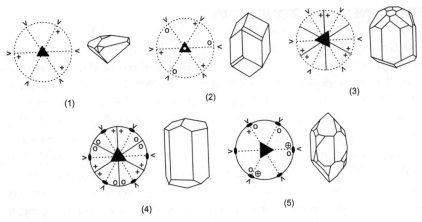

Fig. 2.23 Crystal classes of trigonal system— 3 or $\bar{3}$ axis form the vertical axis: (1) Trigonal hemimorphic [3] magnesium sulphate (2) Rhombohedral [$\bar{3}$] dioptase CuSiO$_3$H$_2$O(3) Ditrigonal hemimorphic [3 m] lithium niobate, LiNbO$_3$ (4) Trigonal holosymmetric [$\bar{3}$ m] calcite, CaCO$_3$ (5) Trigonal trapezohedral [32] α-quartz, SiO$_2$

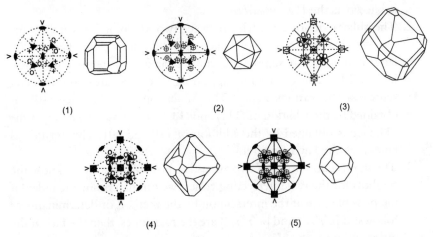

Fig. 2.24 Crystal classes of cubic system: four 3-fold axis are characteristic of this system. Crystallographic axis can coincide with 2-fold; $\bar{4}$ or 4 fold axis: (1) Pentagonal dodecahedral sodium chlorate, NaClO$_3$ (2) Diakisdodecahedral [2/$m\bar{3}$ = $m3$] pyrite FeS$_2$ (3) Hexatetrahedral [$\bar{4}$ 3m] sphalerite, ZnS (4) Pentagonal icositetrahedral [432] cuprite, Cu$_2$O (5) Cubic holosymmetric [$m3m$] a number of minerals like galena, periclase, spinels, garnets, etc. Crystal of copper is shown

under different systems from triclinic to cubic are given in Figs. 2.18 to 2.24. The name of the class, the symmetry elements present and the type of crystals (if known) are also given.

2.6   ZONES AND ZONE LAWS

The faces on a crystal which are parallel to themselves and parallel to a common axis are said to form a zone and are called co-zonal or tauto-zonal faces. The corresponding axis is called the zone axis. In Fig. 2.25, the faces (210), (2$\bar{1}$0), (110), (010), (0$\bar{1}$0) are parallel to themselves and to a common axis-c which is marked as [001]. Similarly, the faces (001), (101), (100), (10$\bar{1}$), (00$\bar{1}$) are again all *co-zonal faces* whose zone axis is [010]. In general, a zone axis is a direction and any direction in a crystal is written in square brackets []. Two symbols [100] and [$\bar{1}$00] indicate the opposing directions of $+a$ and $-a$, respectively. Similarly, (110) and ($\bar{1}$10) are opposing directions. The zone axis need not always coincide with the crystallographic axis. Like the general symbol for a face is (hkl), the general zone symbol is [uvw]. There exist certain general relations with respect to the indices of faces and those of zones.

1. If any face (hkl) is parallel to an axis and belongs to a particular zone [uvw], then it should satisfy the condition $uh + vk + wl = 0$. This relation is known as the *Weiss zone law*.

2. The Miller index of a face lying between two adjoining faces in the same zone is obtained by a simple algebraic addition of the indices of the latter two faces. These points become clear from a stereogram shown in Fig. 2.10. The great circle appearing as circles, arcs and diameter represent the various zones. The Miller index of the face (101) is obtained by the addition of (111) and (1$\bar{1}$1). Likewise, the index of the 111 face is obtained by the addition of 110 and 001. This relation is known as the law of addition.

3. The intersection point of two zones is a possible face. If we know the symbols for the two intersecting zones, we can determine the index of the possible face at the intersection by the method of determinants as follows: If [$u'v'w'$] and [$u''v''w''$] are the two zones, then the face at the intersection point hkl is

$$\begin{array}{ccc} u' \\ u'' \end{array} \begin{array}{ccc} v' \\ v'' \end{array} \begin{array}{ccc} w' \\ w'' \end{array} \begin{array}{ccc} u' \\ u'' \end{array} \begin{array}{ccc} v' \\ v'' \end{array} \begin{array}{ccc} w' \\ w'' \end{array}$$

$$h = v'w'' - v''w'; \; k = w'u'' - w''u'; \; l = u'v'' - u''v'$$

In the same manner, if the indices of two parallel (non-coplanar) faces are ($h_1 k_1 l_1$) and ($h_2 k_2 l_2$), then the zone symbol for the zone [uvw] to which these faces belong can be determined (see Fig. 2.25).

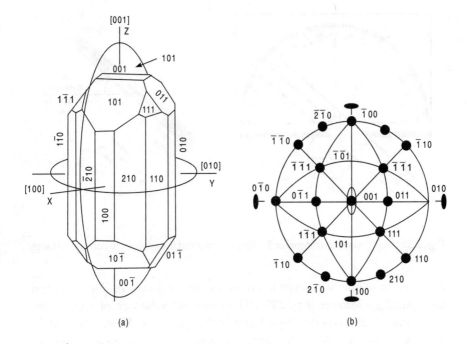

Fig. 2.25   (a) A crystal of lead sulphate (b) Its stereographic projection to show the co-zonal faces and the zone symbol in square brackets

$$\begin{array}{ccc|ccc}h_1 & k_1 & l_1 & h_1 & k_1 & l_1 \\ h_2 & k_2 & l_2 & h_2 & k_2 & l_2\end{array}$$

$$u = k_1 l_2 - k_2 l_1; \; v = l_1 h_2 - l_2 h_1; \; w = h_1 k_2 - h_2 k_1$$

2.7   ANGULAR MEASUREMENTS AND MATHEMATICAL RELATIONSHIPS

As mentioned in the introductory Section 2.1 of this chapter, a constancy is always observed in the measurement of the *interfacial angle*. This feature has been extensively made use of by earlier crystallographers to determine the symmetry and hence, the point group of any given crystal. This is particularly highlighted in what is known as Steno's law or the law of *constancy of the interfacial angle*. This law states that the angle measured between a given pair of faces of specific Miller Indices will be the same for all crystals of the same substance. Due attention was, therefore, given for the accurate measurements of interfacial angles and appropriate instruments have been developed. The

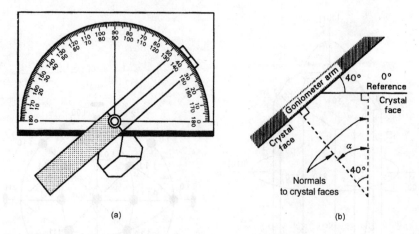

(a)   (b)

Fig. 2.26   (a) Contact goniometer (b) Measurement of the angle between crystal faces

most common ones used even today are the *contact goniometers* (Fig. 2.26) and the optical goniometer (Fig. 2.27). The measured values on plotting in the stereograms will reveal the dispositions of the symmetry elements in relation to the chosen crystallographic axis. This approach to the measurement of interfacial angle and their plotting on the stereograms has been very useful in assigning the point group as well as identifying ill-developed crystals.

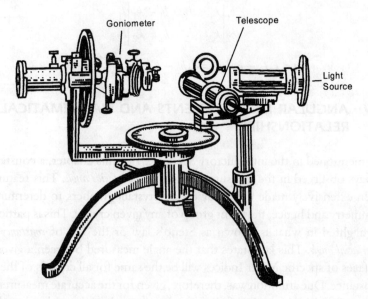

Fig. 2.27 A two- circle goniometer

2.7.1 Measurement of Interfacial Angles

One of the earliest tools in the hands of crystallographers was a contact goniometer, which was invented in 1780 by Carangeot. Figure 2.26 is self-explanatory with regard to its use for the measurement of interfacial angles. The interfacial angle is the angle between the normals drawn to the two faces from the centre of the crystal. Here, the supplement of the angle read on the contact goniometer is considered to form the interfacial angle. The reflecting or optical goniometer was first developed by Wollaston in 1809, and its subsequently improved versions have been most effective in accurately measuring the external forms of crystals and their interfacial angle. A typical *two-circle goniometer* is shown in Fig. 2.27. The crystal is mounted on the goniometer which has a rotational as well as a tilting motion. The light incident on the reflecting plane of the crystal is read by the telescope which is mounted on a circular graduated scale. This is a simple and useful tool for measuring the interfacial angle on very tiny crystals which is invisible to the naked eye.

2.7.2 Some Mathematical Relations

Making use of the measured interfacial angle and the principles of spherical trigonometry, one can calculate the angles between the different poles of faces and determine the axial ratios. Here, the spherical triangles are drawn on the surface of a sphere by the intersection of the zones. As one may recollect from Section 1.6.4, the spherical triangles are different from the planar triangles in that all the six parts of the spherical triangles are angular measurements expressed in degrees. The transferred great circles on the stereographic projection also bear the same relations. In Fig. 2.28, the triangle ABC on the spherical surfaces is projected on the stereographic plane as a triangle A'BC. The trigonometrical relations that can be derived are the following:

a) the determination of one side in terms of the other sides of the spherical triangle and their included angles. In Fig. 2.28 the included angles are

$$\cos c = \cos a \cos b + \sin a \sin b . \cos C$$
$$\text{or } \cos C = (\cos c - \cos a \cos b) / \sin a \sin b, \qquad (2.1)$$

b) to determine an angle in terms of the three sides.

$$\tan A / 2 = \sqrt{\frac{[\sin(s-b)\sin(s-c)]}{\sin s \cdot \sin(s-a)}} \qquad \text{where } s = a + b + c / 2, \quad (2.2)$$

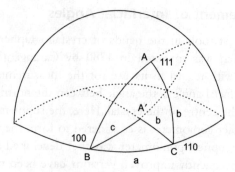

Fig. 2.28   Spherical triangle ABC on a sphere transferred on to an equatorial plane A'BC.

c) to determine the side of spherical triangles in terms of any two angles and a side, or to determine an angle in terms of two sides and an angle not included by them.

$$\sin a / \sin A = \sin b / \sin B . \sin c / \sin C. \qquad (2.3)$$

2.7.3 The Solution of a Napierian Triangle

A spherical triangle where one of the parts is 90° can be used to calculate the unknown angle, if any two of the remaining parts are known. The device named after Napier is as follows. The parts of the triangle are numbered anticlockwise starting from the right angle. A five-component figure is also drawn, where the horizontal line to the right hand side represents the right angle. The values of the angle are marked as in Fig. 2.29. In this figure the sine of the middle part = product of tangent of adjacent parts or product of cosine of the opposite parts. Let us take an example: in Fig. 2.30a, if we chose the triangle formed between poles (111), (110) and (100), and the angles 110^100 = 34°24' and 111^110 = 54°, see Fig. 2.30b and c in order to find the angle between (111)^(100)

$$\sin(90°-c) = \cos 54°\cdot\cos 34°24'$$

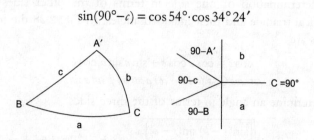

Fig. 2.29   Napier's figure for solution of a right-angled spherical triangle (see text).

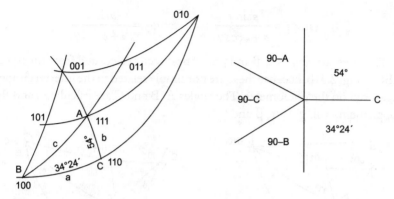

Fig. 2.30 Calculation of the interfacial angle using Napier's device (see text).

2.7.4 The Calculation of the Axial Ratio and Axial Angles

The axial ratio in different crystal systems can be determined by the following relations:

Tetragonal system: $c/a = \tan(001^\wedge 011)$

Orthorhombic system: $a/b = \tan(100^\wedge 110)$; $c/b = \tan(001^\wedge 011)$; $\tan(001^\wedge 101) = c/a$

Monoclinic system:

$$a^\wedge c = \beta = 180 - (\phi_3 + \phi_4) = 180 - [(100^\wedge 001)] \text{ (Fig. 2.31(a))}$$

$$c/a = \sin\phi_3 / \sin\phi_4 = \sin(001)^\wedge(101) / \sin(100)^\wedge(101) \text{ (Fig. 2.31a)}$$

$$\frac{a}{b} = \frac{\sin\phi_1}{\sin\phi_2} = \tan\phi_1 \qquad \frac{c}{b} = \frac{\sin\phi_6}{\sin\phi_5} = \tan\phi_6$$

The other useful relationship in monoclinic systems is:

$$\tan(100^\wedge 110) = \frac{a\sin\beta}{b} \text{ and } \tan 001^\wedge 011 = \frac{c\sin\beta}{b}$$

Similar calculations of the axial ratios for other systems particularly the triclinic system is still more complex owing to six variables involved, namely the three unequal axes and the different included angles. From Fig. 2.31b, various auxiliary angles are related to the *a*, *b*, *c* and interaxial angles by the following relations:

$$\tan\varphi_4 = \frac{a\sin\beta}{c + a\cos\beta} \qquad \tan\varphi_3 = \frac{c\sin\beta}{a + c\cos\beta}$$

$$\tan\varphi_5 = \frac{b\sin\alpha}{c + b\cos\alpha} \qquad \tan\varphi_6 = \frac{c\sin\alpha}{b + c\cos\alpha}$$

$$\tan\varphi_1 = \frac{a\sin\gamma}{b + c\cos\gamma} \qquad \tan\varphi_2 = \frac{b\sin\gamma}{a + b\cos\gamma}$$

The interaxial angles α, β and γ must be directly calculated from the triangle ABC in Fig. 2.31b, because these are not simply related to the interfacial angles measured on the goniometer. The angles A, B and C are calculated and their supplements will give α, β and γ.

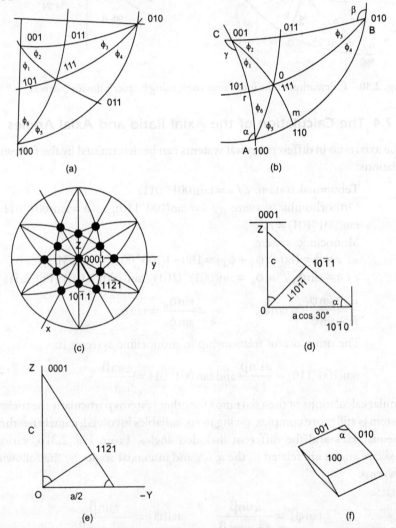

Fig. 2.31 (a) A portion of the stereographic projection of a monoclinic crystal (b) Triclinic crystal $(\alpha \ne \beta \ne \gamma \ne 90°)$ (c) Position of poles (0001) (10$\overline{1}$1) and (11$\overline{2}$1) required to calculate the c/a ratio from Fig. 2.31d (d) Trigonometric relation between c/a ratio and the interfacial angle (0001) and (10$\overline{1}$1) (e) Trigonometric relation to calculate 0001^ 11$\overline{2}$1 in trigonal and hexagonal system (f) The angles α between rhombohedron edges in a trigonal system

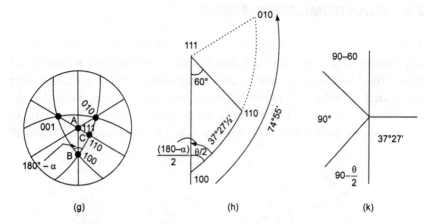

(g)   (h)   (k)

Fig. 2.31 (g) The spherical triangle ABC on a projection of a trigonal crystal which can be used to calculate the angle α of a trigonal crystal by the Napier solution shown in (h) & (k).

If the faces, such as (110), (100), (101), (011), (010) are present in the crystal, then

$$\frac{a}{b} = \frac{\sin 110^{\wedge}100.\sin 010^{\wedge}001}{\sin 010^{\wedge}110.\sin 100^{\wedge}001}$$

In the hexagonal system, the relationship $\tan 0001^{\wedge}10\bar{1}1 = c/a \cos 30°$ is apparent from Fig. 3.31(c) and (d).

Yet another useful relationship in the hexagonal system for the calculation of c/a is $1/2 \tan 0001 ^{\wedge} 11\bar{2}1$, if the position of $11\bar{2}1$ can be calculated from the relation $\tan 0001 ^{\wedge} 11\bar{2}1 = c/\frac{a}{2}$ (see Fig. 2.31e). These relations hold good even for the trigonal system for the calculation of the axial ratio but the angle has to be calculated using the Napier solution. In Fig. 2.31f, the zones 001-100 and 100-010 are normal to the rhombohedron edges which include the angle α. We need to get this angle between these zones on a stereogram (Fig. 2.31g) to get the angle α. If the interfacial angle of the rhombohedron $100^{\wedge}010$ or $100^{\wedge}011$ is measured with a goniometer, then α can be obtained by using Napier's rule on the triangle ABC in Fig. 2.31g. Here, it will give one half of the 180–α, from which the required angle α can be obtained. A solution for α in a calcite crystal whose cleavage angle $100 ^{\wedge} 010 = 74°54'$ is illustrated in Fig. 2.31(h & k). In Fig. 2.31(k), $\cos 60° = \sin\frac{\theta}{2}\cdot\cos 37°27'$ where $\frac{\theta}{2} = 39°2'$ $\alpha = 101°55'$.

2.8   ENANTIOMORPHIC FORMS

There are crystals such as quartz (class 32) which exist in two different dispositions of faces in such a way that the two individual parts are not superimposable. One crystal may be a mirror image of the other (Fig. 2.32), as they are related by right- to left-handed symmetry. Such crystals are said to be enantiomorphic forms. These crystals do not have a plane of symmetry, an inversion axis, or a centre of symmetry. The crystal classes in which such

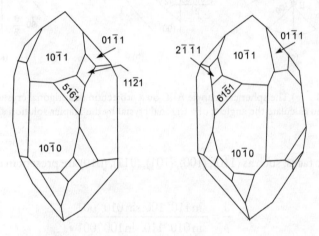

Fig. 2.32   A pair of enantiomorphic crystals of quartz.

enantiomorphic forms are found are 1, 2, 3, 4, 6, 23, 222, 32, 422, 622 and 432. If one observes two enantiomorphic quartz crystals more closely, the faces in the left-handed crystal, i.e., the $(21\bar{1}1)$ and $(61\bar{5}1)$ faces are replaced by the $(11\bar{2}1)$ and $(51\bar{6}1)$ faces, respectively, in the right-handed crystal. The other forms remain the same.

2.9   DEVIATION FROM THE IDEAL GEOMETRY

Like in the world of living things, crystals also deviate from their ideal geometry during their growth. This deviation is related to the deviation in the atomic arrangement in the ideality. As a result, the symmetry of a good number of crystals is not comparable to the 32 point groups. Composite crystals grow in close proximity. When two crystals or two parts of a crystal are related to one another in a crystallographic manner, or through a symmetry operation of reflection or rotation, they are called twinned crystals. The nature of the relation is often expressed in terms of a twin law. The orientation of one individual of a twin crystal is sometimes related to the other individual as though there is a rotation by $360°/n$, where n is most commonly $= 2$. The axis around which such a rotation is seen is called a twin axis (TA). It is a row of lattice points in the structure. When the orientation of two

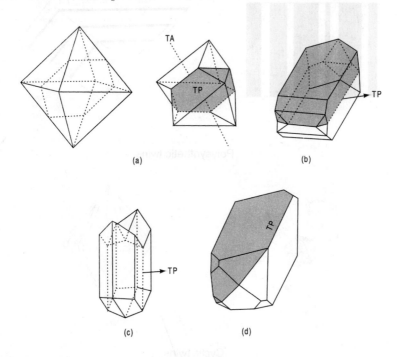

(a)   (b)

(c)   (d)

Contact twins

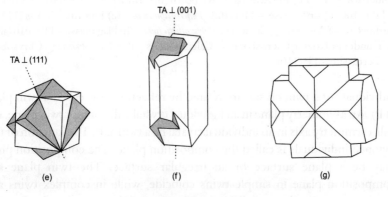

(e)   (f)   (g)

Penetration twins

(continues...)

(... contd)

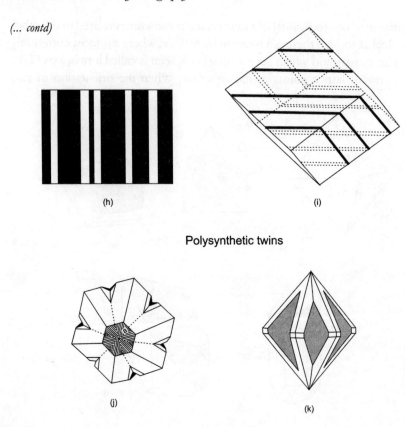

(h)   (i)

Polysynthetic twins

(j)   (k)

Cyclic twins

Fig. 2.33   Some typical types of twins commonly seen in minerals: contact twins (a)
Spinel twin plane(TP) and twin axis (TA) shown (b) Orthoclase–TP is (001) (c) Gypsum
– TP is 100 (d) orthoclase – TP is 021. *Penetration twins* : (e) Fluorite – TA is (111) (f)
Orthoclase–TA is (001) (g) Staurolite. *Polysynthetic twins* : (h) Plagioclase–TP is 010 (albeit
twin under polarizing microscope) (i) Calcite - TP is (001). *Cyclic twins* : (j) Chrysoberyl
(k) Cerrussite – TP (110).

individuals of a twin crystal are related by reflection, it is called a twin plane
(TP). An asymmetry plane in an individual crystal will not be a twin plane, nor
will a symmetry axis in an individual crystal be a twin axis. The junction of two
twinned individuals is called the composition plane. The composition plane
may be a plane surface or an irregular surface. The twin plane and
composition plane in simple twins coincide, while in complex twins and
penetration twins they are different. There are different types of twins such
as simple twins, multiple twins, contact twins, penetration twins, cyclic twins,
etc

Internal Symmetry and Crystal Lattice

3.1 INTRODUCTION

Early crystallographers always thought that external geometry arose out of a regularity in the arrangement of atoms or molecules within the crystal. Although the concept of atoms and molecules came much later, the thinking of the early crystallographers was directed in terms of what are known as the building blocks. The idea originated from the observation by Huygens that a calcite crystal (of trigonal system) when parted, breaks into tiny rhombic prisms of the same geometry. One can similarly observe that a halite crystal parts into tiny cubes. Thus, a large crystal was thought to be built up of such tiny invisible individual *building blocks* (Fig. 3.1a). This concept was further developed by Haüy who proposed that the building blocks be replaced by *spheres* which, in turn, are thought to represent atoms or molecules (Fig. 3.1b). Soon after the discovery of X-rays by Röentgen and the diffraction of X-rays by crystals discovered by Von Laue, it was demonstrated that crystals are equivalent to 3-dimensional gratings made up of atoms or molecules. Subsequent development in X-ray crystallography fully supported the orderly arrangement of atoms or molecules in crystals in 3 dimensions. This chapter deals with the early stages in the development of the concepts of internal symmetry and crystal structure. This, in turn, leads us to the concept of space lattices.

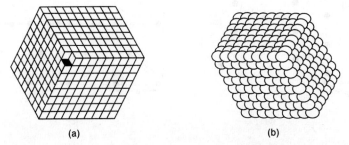

(a) (b)

Fig. 3.1 (a) The concept of building blocks to explain smooth surfaces and geometrical forms in crystals (b) Haüy's model of closely packed hard spheres represent the building blocks.

3.2   ARRAY OF OBJECTS AND TRANSLATIONAL SYMMETRY

If a set of identical objects are arranged in space at regular intervals, we generate an array of objects. If the objects are arranged in a line at regular spacings, it is an 1-dimensional array as in Fig. 3.2a, where the repeating object is a flower. The flowers 1, 2 and 3 are interrelated by a type of symmetry called the *translational symmetry*. Flower 1 is related to flower 2 by a *unit translation*. If the translation is carried out along the *a*-axis, the unit translation is '*a*'; or if the direction is considered as the *b*-axis then the unit is '*b*'. Thus, the distance through which flower 1 is moved to generate flower 2 is a vector quantity and is called an *a*-, *b*- or *c-vector unit of translation*. If the flowers are arranged as in Fig. 3.2b, it is called a 2-dimensional planar array. The translational vector on the *a*-and *b*-axes may or may not be the same. Figure 3.1c represents such repetitions in 3 dimensions of *a*-, *b*- and *c*-vectors. The 3-dimensional array, so generated, is by the repetition of a 2-dimensional planar array with a unit *c*-vector. In order to conceive the nature of the arrangement as a whole rather than as an individual object (a flower in the

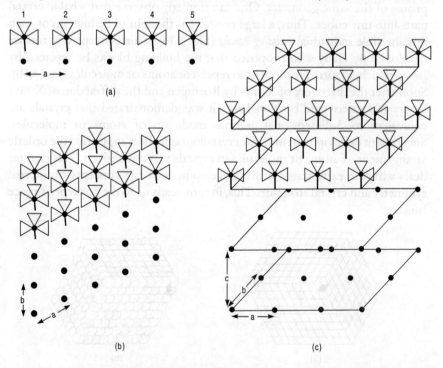

Fig. 3.2   (a) The 1-dimensional periodicity of objects—Array of lattice (b) 2-dimensional periodicity—Planar lattice (c) 3-dimensional periodicty of objects in space—Space lattice.

present case), the objects are replaced by points (also called *identi-points*). An array of identi-points is called the *point lattice*. Thus, Figs. 3.2a to 3.2c are called 1-, 2- and 3-dimensional lattices, respectively. Alternatively, a 1-dimensional lattice is called a lattice row or a line lattice. The 2-dimensional lattice is called a planar lattice or a net. The 3-dimensional lattice is called a space lattice. In an actual crystal lattice, a point does not necessarily represent only one atom or one molecule. It could be a group of atoms or molecules, where every point correspondingly represents the same group of atoms repeating.

3.3 PLANAR LATTICE TYPES

As mentioned earlier, in a line lattice, the identi-points are equally spaced along a line and only one line lattice type is possible.

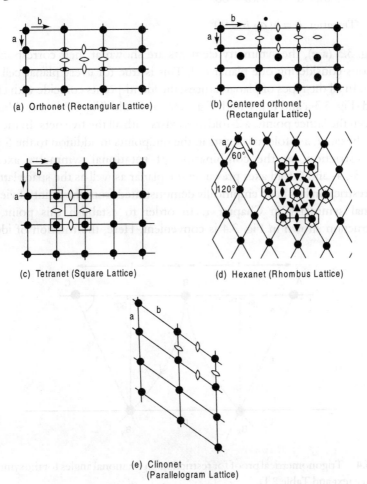

(a) Orthonet (Rectangular Lattice)

(b) Centered orthonet (Rectangular Lattice)

(c) Tetranet (Square Lattice)

(d) Hexanet (Rhombus Lattice)

(e) Clinonet (Parallelogram Lattice)

Fig. 3.3 The five basic planar lattice types.

A *planar lattice* or a net consists of two-line lattices intersecting one another. The angle of intersection of the two-line lattice can differ. In addition, the unit of vectors (*a* and *b*) on these two interlocking lattices may be identical or different. Accordingly, only the following types of planar lattices are possible [Fig. 3.3(a–e)].

1.   Clinonet $a \neq b, a \wedge b \neq 90°$

2.   Orthonet $a \neq b, a \wedge b = 90°$

3.   Centred orthonet $a \neq b, a \wedge b = 90°$ (but because of the point at the centre of the net, *b* is the shortest translation distance next to *a*)

4.   Hexanet $a = b, a \wedge b = 60°$

5.   Tetranet $a = b, a \wedge b = 90°$

In Fig. 3.3 (a-e), the symmetry elements are shown at their corresponding locations considering one planar cell. This is true for every planar cell of a lattice. In all the types of planar lattices, the identi-points coincide with either 2-fold (Fig. 3.3(a and b)), 4-fold (Fig. 3.3c) or 6-fold (Fig. 3.3d) axes. Midway between the lattice points, a 2-fold axis exists with all the five nets. In the case of the hexanet, a 3-fold axis exists in the midpoints in addition to the 6-fold. The restriction on the combination of rotational symmetry axes to 1-, 2-, 3-, 4- and 6-fold axes prevail in the planar as well as the space lattices. This restriction has been empirically demonstrated from the point of view of external symmetry in Chapter I. In order to establish this point, the construction shown in Fig. 3.4 is convenient. Here, the rotation of identi-

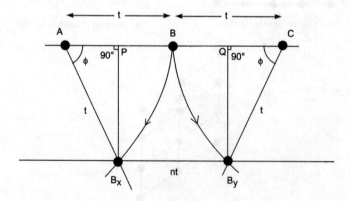

Fig. 3.4   Trigonometrical proof for restriction on rotational angles for the symmetry axis (see text and Table 2.1).

Table 3.1 *The possible values of the repeat angle which restrict the possible axes of symmetry in a planar lattice of periodic repetition, obtained from the relation* $\cos \phi = 1 - n/2$ *from Fig 3.4.*

n Integral multiple of translation vector t	0	1	2	3	4
ϕ Possible value of repeated angle	360°	60°	90°	120°	180°
Rotation axis	1 or $\bar{1}$	6 or $\bar{3}$	4 or $\bar{4}$	3 or $\bar{3}$	2 or $\bar{2}$

points B in the clockwise direction from the point A or anticlockwise direction from C by the same angle ϕ, should lead to two new lattice points B_x and B_y whose distance should be an integral multiple of the translational vector 't', i.e., 'nt'. The distance between A and C is $2t$, and the distance between B_x and B_y is an integral multiple of 't', i.e., 'nt' where $n = 1, 2, 3 \ldots$

$$\text{The distance } 2t = AP + CQ + PQ = t\cos\phi + t\cos\phi + nt \quad (3.1)$$
$$= 2t\cos\phi + nt.$$

Hence, the Eq.(3.1) can be rewritten in terms of $\cos\phi$, namely $\cos\phi = 1 - n/2$. Since cosine is restricted to values between -1 and $+1$, only the values of $n = 0, 1, 2, 3$ or 4 offer a valid solution to this expression of $\cos\phi$. Table 3.1 shows the permissible values of the repeat angles ϕ which are related to the value of n. This restricts the axis of symmetry at every lattice point in a plane lattice (with periodic repetition of points) to being one of the ten indicated in Table 3.1.

3.4 THE POSSIBLE SPACE LATTICES

As mentioned earlier, the stacking of planar lattices leads to the 3-dimensional space lattices. Such stackings could only give rise to 14 different types of space lattices, which were first demonstrated by the French crystallographer Auguste Bravais, and hence called the 14 Bravais lattices. The planar lattices should be parallel when stacked. In such a stacking process, the resulting space lattice type depends on: a) the symmetry of the planar lattice being stacked, b) the length of the stacking vector, and c) the angle of the stacking vector to the net.

3.4.1 Cubic Space Lattices

The three possible types of lattices based on the tetranet (square lattice) are shown in Fig. 3.5(a–c). In Fig. 3.5a, the stacking vector is at 90^0 to the tetranet.

Further, the length of the stacking vector is equal to the planar vector. One smallest possible unit of such a space lattice is called a unit cell, which in this case has eight identi-points at eight corners of the cube. It is called a primitive cubic cell. The lattice is called the primitive cubic lattice (shortened to cubic-P). In Fig. 3.5b, the translational vector is at $54°44'$ to the net and the stacking length is $0.866a$. This can also be visualized as arising as a result of the parallel sliding of net-2 in relation to net-1, or net-3 in relation to net-2, and so on. Because of the above mentioned angles and the lengths of the vectors, the lattice points in net-3 will come above those of net-1, and so on. That is, the points on the alternate stacking net are at a distance of a. Because of this, the unit cell of this lattice type is a cube with an additional point at the body centre. This is called a body-centred cubic lattice, in short, termed as cubic-I (originating from the German word 'innenzentrierte'). Figure 3.5c illustrates the third possibility of the space lattice type by the stacking of the tetranet. Here, the tetranet is modified such that there is an identi-point at the centre of each square. The centred tetranet by itself is not a primary planar lattice type. Considering that the modified tetranet is stacked, with the stacking angle of $45°$ we obtain the vector length a''. The within net vector a' is at $60°$ to vector a'. But a' equals a''. In this mode of stacking, the identi-points of alternate layers will lie one above the other on the perpendicular axis. The distance between two identi-points of alternate layers is also equal to a. The resulting symmetry of the unit cell in this arrangement has all the symmetry elements of a cube. At the same time, the unit cell differs from primitive and body centred cells in having identi-points on all the face centres. Hence, it is called a face-centred cubic lattice, in short, termed as cubic-F. In all the above three cases the point group remains the same, namely $4/m\ \overline{3}\ 2/m$, but the corresponding space group symbols will be P $4/m\ \overline{3}\ 2/m$, I $4/m\ \overline{3}\ 3/m$ or F $4/m\ \overline{3}\ 3/m$. These are, in short, termed as P$m3m$, I$m3m$ and F$m3m$, representing the three different space lattices under the cubic system. The primitive, body or face-centred unit cells also differ in their content of identi-points. In the primitive cell, each corner point is shared by eight unit cells, because eight cubes surround a given point in three-dimension. In effect, the total number of identi-points belonging to one cell is $8 \times 1/8 = 1$. Hence, it is called a primitive unit cell. Similarly, a body-centred unit cell is doubly primitive because, in effect, it contains two identi-points. Likewise, a face-centred unit cell is quadruply primitive because it contains four identi-points, because each of the lattice points at the face centre is shared by two adjoining unit cells.

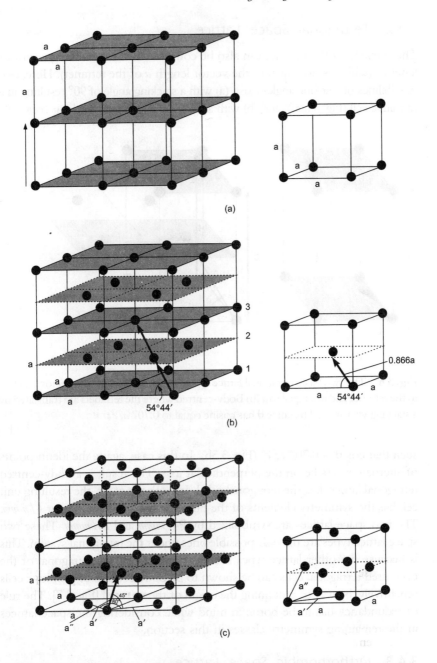

Fig. 3.5 The three cubic lattice types built with a tetranet: (a) primitive stacking at 90° to the tetranet, with the stacking vector equal to *a*. (b) Body-centred stacking at 54°44′ to the vector *a* and length equal to 0.866 *a* (c) Face-centred stacking at 45° to the tetranet, but at 60° to each of the within net vectors labelled a'. $a^\wedge a'' = 45°$, $a'^\wedge a' = 60°$.

3.4.2 Tetragonal Space Lattice

The stacking of the tetranet can also be considered with a stacking vector of length c (which is not equal to the vector length a of the tetranet). Here, two possibilities of stacking angles exist: (a) with a stacking angle of 90° resulting in a primitive cell (Fig. 3.6(a)), and (b) with a stacking angle θ and stacking vector c',

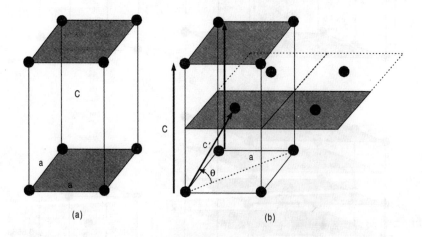

(a)   (b)

Fig. 3.6 The two basic tetragonal lattice types: (a) Primitive stacking vector C at 90° to the tetranet and not equal to a (b) Body-centred where the tetranets are translated by a stacking vector C'. The angle θ has cosine equal to 0.707 a/c'.

such that $\cos θ = 0.707\,a/c'$ (Fig. 3.5b). In this case, again, the identi-points of alternate layers lie on the perpendicular axis. The result is a body-centred tetragonal lattice, i.e., the tetragonal cell-I. In both the cases, the resulting unit cell has the symmetry elements of the point group $4/m\,2/m\,2/m$ or $4/mmm$. The two space lattices are symbolized by P4/mmm and I4/mmm. These two space lattice types are the only possible ones under the tetragonal system. This is because the other lattice types such as the face-centred tetragonal or the C-centred tetragonal cells can be shown to be body-centred or primitive cells repetitively, by way of redefining the origin of the unit cell [Fig. 3.7]. The rule of redundancy has to be borne in mind while considering the space lattices in the remaining symmetry classes in this section.

3.4.3 Orthorhombic Space Lattice

The stacking of the orthonet directly over one another with a stacking vector angle of 90° and a stacking vector length c (where $a = b = c$), results in a primitive orthorhombic space lattice (Fig. 3.8). If the orthonets are stacked

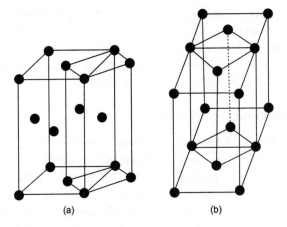

Fig. 3.7 (a) The face-centred tetragonal lattice been shown to be defined by a body-centred lattice (b) Likewise, the *c*-centred tetragonal lattice can be defined with a primitive lattice.

at an angle, it will give rise to a body-centred orthorhombic lattice. The stacking of centred orthonets with stacking angles of 90°, yields a lattice whose cells will have identi-points at the centre of two parallel faces. This is called the *C*-centred orthorhombic lattice (Fig. 3.8c). Depending upon the direction of stacking, we can have *a*-face-centred or *b*-face-centred lattices. However, *a*-, *b*-, or *c*-face-centred lattices represent the same type with a

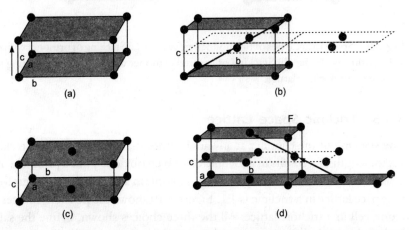

Fig. 3.8 Space lattice types of the orthorhombic system resulting from the stacking of orthonets: a) Primitive lattice b) Body-centred lattice c) C-face centred lattice (d) Face-centred lattice.

difference in orientation. We can also consider the stacking of centred orthonets at an angle (Fig. 3.8d). This results in a face-centred orthorhombic space lattice. In brief, the orthorhombic point group with the symmetry 2/ m 2/ m 2/ m (or mmm) can have four orthorhombic space lattices, symbolized as Pmmm, Immm, Fmmm and C(or A or B) mmm.

3.4.4 Monoclinic Space Lattice

The stacking of the orthonet with a vector at an angle of 90° to the b-direction and other than 90° to the a-direction will result in primitive monoclinic space lattices. The stacking of centred orthonets at an angle β, which is not equal to 90°, also results in a monoclinic lattice, but with C-centred monoclinic cells. Thus, there will be two space lattices under the monoclinic system, P2/m and C2/m (Fig. 3.9 (a and b)). A face-centred or body-centred monoclinic cell cannot exist as it generates a large number of symmetry elements corresponding to an orthorhombic cell which has already been defined.

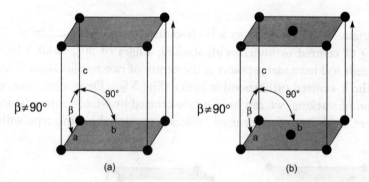

Fig. 3.9   Space lattice types of the monoclinic system: (a) Stacking of orthonet with $c^\wedge a$ ≠ 90°, primitive lattice (b) Stacking of centred orthonet with $c^\wedge a$ ≠ 90°, resulting in a C-centred monoclinic lattice.

3.4.5 Triclinic Space Lattice

The stacking of the clinonet at any angle other than 90°, but with a stacking vector length of c, results in a triclinic cell which obviously has a primitive unit cell. The maximum symmetry elements possible in such a case is $\bar{1}$. Therefore, the space lattice in a triclinic is P$\bar{1}$. Figure 3.10 shows the possible choices of a unit cell in a triclinic lattice. All the three choices shown, define the same lattices, but with different repeat distance along the a, b and c directions and with different angles: α, β and γ. The convention is to choose the cell with the shortest repeat distance as the unit cell.

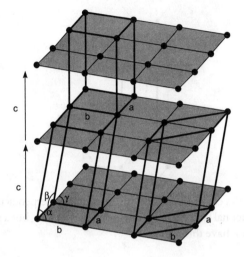

Fig. 3.10 A space lattice in a triclinic system obtained by stacking of clinonet. The thick lines define the choice of the unit cell.

3.4.6 Hexagonal and Rhombohedral Space Lattices

The stacking of the hexanet one over the other with a stacking vector $c (\neq a)$ at an angle of 90° results in a new space lattice having all the symmetry elements corresponding to $6/m\ 2/m\ 2/m$. This hexagonal prism is triply primitive because 1/6th of the corner identi-points belong to the unit cell, so that totally $1/6 \times 12 = 2$ (six each on the top and bottom of the cell together with $1/2 \times 2 = 1$ identi-point from the c-face centred). This makes three identi-points belonging to this cell. Alternatively, a primitive cell can be considered for the hexagonal lattice, taking only 1/3rd of the hexagonal cell, which is a rhombic prism, as shown in Fig. 3.11. Therefore, the *hexagonal lattice* is invariably primitive, which can be symbolized as P6/*mmm*.

A second way of stacking the hexanets to form a new space lattice is shown in Fig. 3.12. Here, the stacking of the second set is at an angle other than 90°, and the stacking vector is the distance between any two identi-points of net-1 and net-2, which is the edge length of the rhombohedron. The subsequent layers 3, 4, etc., are also moved in such a way that the stacking vector angle α remains the same. The identi-points of net-1 and net-4 directly lie one above the other, so that a modified hexagonal unit cell can be drawn between layers 1 and 4 (Fig. 3.12). In this arrangement there is a relation between the identi-points of net-2 and net-3, i.e., if the identi-points on net-2 lie on one set of alternate 3-fold axes of the hexanet (marked 2 in Fig. 3.12b), then the identi-points of layer-3 will coincide with the other set of alternating

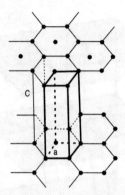

Fig. 3.11   A primitive cell of a hexagonal lattice formed by perpendicular stacking of the hexanet. The hexagonal lattice has a point group of 6/mmm, but a primitive unit cell (bold lines) does not have this symmetry.

3-fold axis (marked 3 in Fig. 3.12b). Such a lattice arrangement lowers the symmetry of the point group $\overline{3}2/m$ to $\overline{3}m$. The smallest primitive cell that retains this symmetry can now be outlined in terms of a rhombohedron. The two parameters that define the size of the rhombohedral cell are 'a', the edge length and α, the acute angle between any two edges. The point where the edges with acute angles meet, marks the emergence of the 3-fold axis. The lattice is symbolized as R and the space lattice is symbolized as $R\overline{3}2/m$.

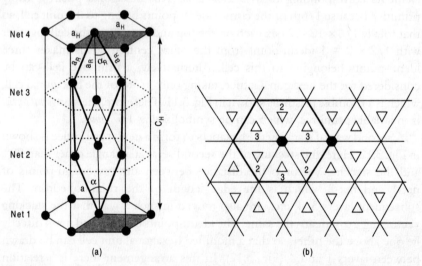

Fig. 3.12   (a) An alternative way to stack hexanets when the unit cell will have the rhombohedral symmetry $R(\overline{3}2/m)$. Here the stacking vector is not perpendicular, but coincides with one of the edges of rhombohedron. (b) Relation between identi-points of net 2 and 3 on plan view. Note the two types of 3-fold sites

In effect, two alternate choices of unit cell are possible for a rhombohedral lattice. As mentioned earlier, the first choice is in terms of a rhombohedral cell with a_R and α_R as two parameters. The second choice is in terms of the rhombic right prism of the height c_H. The base of this prism is a rhombus with a side a_H and an acute angle of 60° (Fig. 3.11). The two sets of parameters are interconvertible with the following relations:

From rhombohedral to hexagonal $a_H = 2a_R \sin\dfrac{\alpha_R}{2}$

$$c_H = \sqrt{9a_R^2 - 3a_H^2}$$

From hexagonal to rhombohedral $a_R = 1/3 \cdot \sqrt{3a_H^2 + c_H^2}$

$$\sin\frac{\alpha_R}{2} = \frac{3a_H}{2\sqrt{3a_H^2 + c_H^2}}$$

The above discussion illustrates that the *rhombohedral lattice* can be considered a part of the hexagonal lattice. In the rhombohedral lattice, one can also consider the body-centred as well as face-centred unit cells, since the 3-fold symmetry permits such a possibility. However, as shown in Fig. 3.13, this produces no new lattice arrangement since the reorientation of the unit cell results, again, in a primitive rhombohedral cell in both the cases.

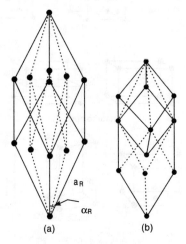

Fig. 3.13 (a) A face-centred rhombohedral lattice will result in a primitive rhombohedral lattice. (b) Likewise, consideration of a body-centred rhombohedral lattice also produces a primitive rhombohedral cell.

3.4.7 The 14 Bravais Lattices

The possible space lattices discussed in the previous section lead us to the conclusion that only 14 space lattice types are possible, which are called the

Bravais lattices. These are summarized in Table 3.2, and are collectively drawn in Fig. 3.14.

Table 3.2

Crystal System	Lattice Type					Symmetry
	P	I	F	C	R	
Triclinic	×	–	–	–	–	$\bar{1}$
Monoclinic	×	–	–	×	–	$2/m$
Orthorhombic	×	×	×	×	–	mmm
Tetragonal	×	×	–	–	–	$4/mmm$
Hexagonal	×	–	–	–	–	$6/mmm$
Trigonal*	–	–	–	–	×	$\bar{3}2/m$
Isometric	×	×	×	–	–	$m3m$

* Some trigonal crystals possess a hexagonal P lattice, but others have the rhombohedral R-type lattice. For the R-type, the lattice parameters are a_R and α_R [Fig. 3.13a].

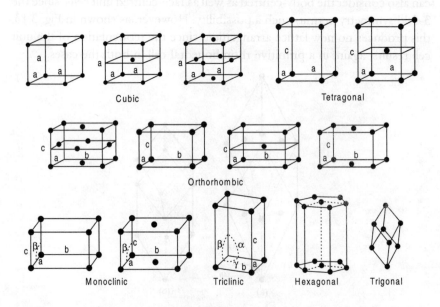

Fig. 3.14   The 14 Bravais lattice types which belongs to seven crystal systems.

3.5   SIMPLE SPACE GROUPS

The total symmetry of a 3-dimensional array is dependent upon two factors: a) the inherent symmetry of the object which is being translated to create an array, and b) the symmetry of the space lattice or the scheme of translation in the array itself. This combined symmetry is generally referred to as a space group. As it is evident from the previous section, the 14 possible space lattices possess the symmetry elements of the holohedral class, i.e., the point group with the highest symmetry elements under each system. However, the space lattices can exist with point groups of lower symmetries under various crystal systems (hemihedral, tetrahedral class). For example, the cubic space lattice need not only have $4/m\,\bar{3}\,2/m\,(m\,3m)$ symmetry, but can also have $\bar{4}\,3m$, 432, $2/m\,\bar{3}\,(m3)$ or 23 symmetries as well. Thus, in the cubic crystal system the P, I and F lattices exists for all the 5 point groups, in turn giving rise to 15 space groups. The decrease in symmetry characters is essentially due to the decrease in the symmetry of the object itself, although they can be translated to form an array with P, I and F types of cubic lattices. However, this has a limitation posed by the lack of symmetry of the object itself. Since a low symmetry object cannot be arranged into a cubic space group, accordingly the system will have to change with a correspondingly lower symmetry class. The space groups generated by combining the permitted space lattice type with the allowed point groups under each system, are called simple space groups (Table 3.3), which are 72 in number.

Table 3.3   *The 72 Simple Space Groups under the Seven Crystal Systems*

System	Simple Space Groups
Triclinic	P1; P$\bar{1}$
Monoclinic	P2; Pm; P2/m C2; Cm; C2/m
Orthorhombic	P222; Pmm 2; Pmmm C222; Cmm 2; Amm 2; Cmmm F222; Fmm 2; Fmmm I222; Imm 2; Immm
Tetragonal	P4; P$\bar{4}$; P4/m; P 422; P 4mm; P$\bar{4}$ 2m, P $\bar{4}$ m2; P 4/mmm I4; I$\bar{4}$; I4/m; I 422; I 4mm; I $\bar{4}$ 2m I $\bar{4}$ m2; I 4/mmm
Trigonal	P 3; P $\bar{3}$; P 312; P 321; P 3m1; P 31m P $\bar{3}$ 1 2/m; P $\bar{3}$ 2/m 1 R 3; R $\bar{3}$; R 32; R 3m; R $\bar{3}$ 2/m
Hexagonal	P 6; P $\bar{6}$; P 6/m; P622, P 6mm; P $\bar{6}$ m2 P $\bar{6}$ 2m; P 6/mmm
Isometric	P 23; P m3; P432; P $\bar{4}$ 3m; P m3m I 23; I m3; I 432; I $\bar{4}$ 3m; I m3m F 23; F m3; F 432; F $\bar{4}$ 3m; F m3m

Three cases of space groups mentioned in Table 3.3 need special attention. The C*mm*2 and A*mm*2 are considered as two separate space groups, though in the Bravais space lattice, the *A*-centred or *C*-centred ones are not distinguished. This is because the 2-fold axis of the object being repeated is perpendicular to the centred face of the lattice in C*mm*2, while it is parallel to it in A*mm*2. This is illustrated in Fig. 3.15 (a and b). In a tetragonal system, P $\overline{4}$2*m* and P $\overline{4}$*m*2 are again listed as independent space groups. Here, the 2-fold axis of the object coincides with the translational vector '*a*' in the case of P $\overline{4}$2*m*, whereas the 2-fold axis is at 45° to the vector '*a*' in the case of P $\overline{4}$*m*2 (Fig. 3.15(c and d)). The space groups P3*m*1 and P31*m* are also listed separately in Table 3.3. Here again, for P3*m*1 (Fig. 3.15e), the mirror plane of the object is perpendicular to the lattice vector '*a*'. In comparison, the mirror

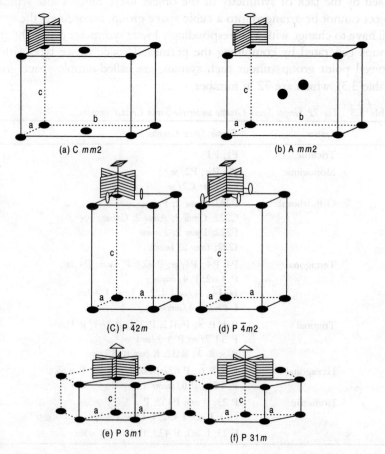

(a) C *mm*2

(b) A *mm*2

(C) P $\overline{4}$2*m*

(d) P $\overline{4}$*m*2

(e) P 3*m*1

(f) P 31*m*

Fig. 3.15   The change in space groups resulting from a change in the location of lattice points with respect to symmetry elements.

planes of the objects are parallel to the lattice vector '*a*' in the case of P 31*m* [Fig. 3.15f].

3.6 INTERNAL SYMMETRY ELEMENTS RELATED TO TRANSLATION : SCREW AXES AND GLIDE PLANES

The basic symmetry elements used to arrive at the 32 point groups, namely 1, 2, 3, 4, 6, $\overline{1}$ (= *i*), $\overline{2}$ (= *m*), $\overline{3}$, $\overline{4}$, $\overline{6}$, are based on external morphological features. On the other hand, the internal symmetry of crystals, accountable in terms of the space lattices, have to also take into consideration the translational symmetry. Thus, combining the rotational operation with the translation brings in new symmetry features known as the screw axes. Similarly, combining the mirror plane with the translation produces the glide plane.

3.6.1 Screw Axes

In order to understand the symmetry operations involved in a screw axis, let us consider a set of identical objects which are asymmetric, namely, a_1, a_2, a_1', a_2', etc. In Fig. 3.16a, the object a_1 is related to a_2 by a rotation of 180°, also a_3 to a_4, a_5 to a_6, and so on. In a 2-fold axis involving rotation and translation symmetry, the pair a_1–a_2 is related a_3–a_4 by a translational vector *t*. Consider the situation in Fig. 3.16b. Here, a_1 is related to a_2 by a rotation of 180° followed by a translation by a vector length of $t/2$, and again by a rotation of a_2 by 180° and a translation by $t/2$ to produce a_3. Thus, the symmetry operation involved is a rotation combined with translation. That is, the objects a_1, a_2, a_3, etc., are related to one another by what is called a 2-fold screw axis. A view from the top of both these arrangements is identical showing only the 2-fold character, while the side-view reveals that this is only an apparent situation. Thus, the external symmetry of both these arrangements give rise only to a 2-fold rotation symmetry. The 2-fold screw axis is expressed by the symbol 2_1 in order to distinguish it from the normal 2-fold axis which has the symbol 2. The symbol 2_1 also means that the translational vector is $t/2$, i.e., 1/2 the total translation. The subscript 1 forms the numerator and the 2-foldness forms the denominator.

The 3-fold axis involving a translational symmetry can generate two more additional symmetry axes as illustrated in Fig. 3.16(c, d and e). In the 3-fold axis, objects a_1, a_2 and a_3 objects are related to one another by a rotation of 120°. The a_1, a_2 and a_3 objects are related to a_1', a_2' and a_3' by a translational symmetry vector *t*. In the case of the screw axis 3_1, the object a_1 is related to

a_2 by a 120° anticlockwise rotation together with the translation by a distance of $t/3$. Repeating this operation relates a_2 to a_3 and a_3 to a_1' by a translation of $t/3$, a_1 is related to a_1' by the translation vector $t/3$. The screw axis 3_2 arises from the 120° clockwise rotation followed by $t/3$ translation. Effectively, 3_1 and 3_2 differ only in the sense of rotation direction. Thus, the axes 3_1 and 3_2

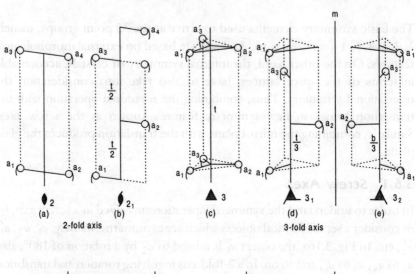

2-fold axis   3-fold axis

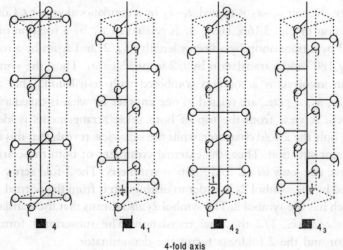

4-fold axis

(continues...)

(...contd)

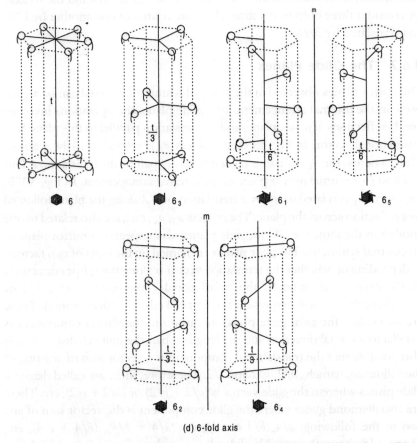

Fig. 3.16 The types of internal symmetry produced through translation '*t*' associated with rotation. The normal rotation axis are also shown: (a) 2-fold axis (b) 3-fold axis (c) 4-fold axis and (d) 6-fold axis. Note the enantiomorphic relation (right- and left-handed screw axis) between 3_1 & 3_2, 4_1 & 4_3, 6_1 & 6_5 and 6_2 & 6_4 axes

are the mirror images of one another. The external morphological features resulting from the difference in the internal arrangement by way of 3_1 and 3_2 only give rise to pairs of enantiomorphous crystals, as in the case of quartz (see Fig. 2.32).

In the 4-fold axis, besides the translation *t*, one can visualise a 90° rotation followed by a *t*/4 translation, leading to 4_1 or 4_3 screw axis depending upon the direction of rotation. It is also possible to consider two pairs of objects related by 180° rotation, and a translation of *t*/2 giving rise to a 4_2 axis. Here again, 4_1 and 4_3 are enantiomorphous. The 6-fold axis involving roto-translation gives rise to, 6_1, 6_2, 6_3, 6_4 and 6_5 axis. Here, 6_1 and 6_5 are enantiomorphous and so also 6_2 and 6_4. In order to understand the 6_3 axis, we consider three objects at a time which are related to one another by 120° and translated by $3t/6$.

3.6.2　The Glide Planes

The mirror planes observed in the external feature of a crystal can be a glide plane in the internal arrangement. In a glide plane, the operations involved are: 1) reflections across a plane, and 2) translation parallel to this plane by a translational vector. In Fig. 3.17a, the object a_1 is related to a_2 by a mirror plane, whereas the a_1-a_2 pair is related to a_3-a_4 pair by the translation t. This will lead to a normal mirror plane in the internal arrangement. In Fig. 3.17b, the object a_1 is related to a_2 by a translation of $t/2$ along the plane followed by a reflection across the plane. The objects a_3, a_4, etc., are also related to one another in the same way. Thus, glide planes also appear as a mirror plane in an external symmetry. The glide planes are named on the basis of two factors: 1) depending on whether the translation vector is along the a, b or c directions of the crystal. Accordingly, the translational component is $a/2$, $b/2$ or $c/2$, where the a, b and c represent the unit cell edges (lattice vectors). These are also called the axial glide planes because the translation component is parallel to a crystal direction and the length is half the unit cell distance. It is also possible that the translational component is a vector sum of any two of the following, namely, $a/2$, $b/2$ and $c/2$. Therefore, these are called diagonal glide planes wherein the glide vector is $(a/2 + b/2)$ or $(b/2 + c/2)$, etc. There are also diamond glides where the glide component is the vector sum of any two of the following: $a/4$, $b/4$ or $c/4$; i.e. $(a/4 + b/4)$, $(b/4 + c/4)$, etc. Figure 3.18 shows the possible glide planes on three faces of an orthorhombic crystal. It may be noted that on the (100) plane the a-glide does not exist. On the (010) plane a b-glide does not exist so also the (001) plane carries no

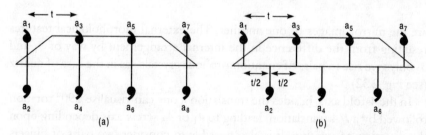

Fig. 3.17　The combination of translation and reflections in internal symmetry: (a) Translation is by distance t (b) Translation is by $t/2$ followed by reflection.

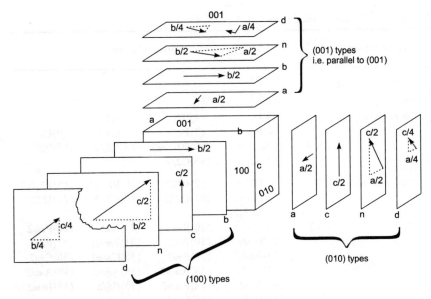

Fig. 3.18   The possible glide planes showing the three perpendicular planes of an ortho-rhombic crystal.

c-glides. The symbols used in the space group notations for the glide planes also follow those given in Fig. 3.18.

3.7   THE 230 SPACE GROUPS

In the previous section, we saw that the combination of the 32 point groups with the Bravais lattices gave rise to the 72 simple space groups. However, the symmetry elements considered there involved a normal rotation axis and normal mirror planes without the translation operation. Considering the fact that the rotation axis can be a screw axis as well, i.e., a 2_1 is equally possible in addition to a 2; or 3_1 and 3_2 are equally possible in addition to 3, and so on. Similarly, in the place of a mirror plane *m*, an *a-, b-, c-, n-* or *d*-glide is possible. Because of these additional possibilities of symmetry operations, more types of combinations leading to newer space groups are possible. Thus, there are a total of 230 such possible space groups. The description of each one of these space groups is too voluminous and hence they are just listed in Table 3.4. Those interested in the descriptive aspects of individual space groups may refer to the International Tables of Crystallography, Vol. 1. The Schoenflies equivalent notations of these 230 space groups are given in Table 3.6. These notations will be of interest to the students of molecular spectroscopy.

Table 3.4   The 230 Space groups

Crystal System	Point Group	Space Groups			
Triclinic	1	*(1)P1			
	$\bar{1}$	(2)P$\bar{1}$			
Monoclinic	2	**(3)P2	(4)P2$_1$	(5)C2	
	m	(6)Pm	(7)Pc	(8)Cm	(9)Cc
	2/m	(10)P2/m	(11)P2$_1$/m	(12)C2/m	(13)P2/c
		(14)P2$_1$/c	(15)C2/c		
Orthorhombic	222	(16)P222	(17)P222$_1$	(18)P2$_1$2$_1$2	(19)P2$_1$2$_1$2$_1$
		(20)C222$_1$	(21)C222	(22)F222	(23)I222
		(24)I2$_1$2$_1$2$_1$			
	mm2	(25)Pmm2	(26)Pmc2$_1$	(27)Pcc2	(28)Pma2
		(29)Pca2$_1$	(30)Pnc2	(31)Pmn2$_1$	(32)Pba2
		(33)Pna2$_1$	(34)Pnn2	(35)Cmm2	(36)Cmc2$_1$
		(37)Ccc2	(38)Amm2	(39)Abm2	(40)Ama2
		(41)Aba2	(42)Fmm2	(43)Fdd2	(44)Imm2
		(45)Iba2	(46)Ima2		
	mmm	(47)Pmmm	(48)Pnnn	(49)Pccm	(50)Pban
		(51)Pmma	(52)Pnna	(53)Pmna	(54)Pcca
		(55)Pbam	(56)Pccn	(57)Pbcm	(58)Pnnm
		(59)Pmmn	(60)Pbcn	(61)Pbca	(62)Pnma
		(63)Cmcm	(64)Cmca	(65)Cmmm	(66)Cccm
		(67)Cmma	(68)Ccca	(69)Fmmm	(70)Fddd
		(71)Immm	(72)Ibam	(73)Ibca	(74)Imma
Tetragonal	4	(75)P4	(76)P4$_1$	(77)P4$_2$	(78)P4$_3$
		(79)I4	(80)I4$_1$		
	$\bar{4}$	(81)P$\bar{4}$	(82)I$\bar{4}$		
	4/m	(83)P4/m	(84)P4$_2$$m$	(85)P4/n	(86)P4$_2$/n
		(87)I4/m	(88)I4$_1$/a		
	422	(89)P422	(90)P42$_1$2	(91)P4$_1$22	(92)P4$_1$2$_1$2
		(93)P4$_2$22	(94)P4$_2$2$_1$2	(95)P4$_3$22	(96)P4$_3$2$_1$2
		(97)I422	(98)I4$_1$22		
	4mm	(99)P4mm	(100)P4bm	(101)P4$_2$$cm$	(102)P4$_2$$nm$
		(103)P4cc	(104)P4nc	(105)P4$_2$$mc$	(106)P4$_2$$bc$
		(107)I4mm	(108)I4cm	(109)I4$_1$$md$	(110)I4$_1$$cd$
	$\bar{4}$2m	(111)P$\bar{4}$2m	(112)P$\bar{4}$2c	(113)P$\bar{4}$2$_1$$m$	(114)P$\bar{4}$2$_1$$c$
		(115)P$\bar{4}$$m$2	(116)P$\bar{4}$$c$2	(117)P$\bar{4}$$b$2	(118)P$\bar{4}$$n$2
		(119)I$\bar{4}$$m$2	(120)I$\bar{4}$$c$2	(121)I$\bar{4}$2m	(122)I$\bar{4}$2d
	4/mmm	(123)P4/mmm	(124)P4/mcc	(125)P4/nbm	(126)P4/nnc
		(127)P4/mbm	(128)P4/mnc	(129)P4/nmm	(130)P4/ncc
		(131)P4$_2$/mmc	(132)P4$_2$/mcm	(133)P4$_2$/nbc	(134)P4$_2$/nnm
		(135)P4$_2$/mbc	(136)P4$_2$/mnm	(137)P4$_2$/nmc	(138)P4$_2$/ncm
		(139)I4/mmm	(140)I4/mcm	(141)I4$_1$/amd	(142)I4$_1$/acd

(continues...)

Table 3.4 The 230 Space groups   *(...contd)*

Crystal System	Point Group	Space Groups			
Trigonal	3	(143) P3$_1$	(144) P3$_1$	(145) P3$_2$	(146) R3
	$\bar{3}$	(147) P$\bar{3}$	(148) R$\bar{3}$		
	32	(149) P312	(150) P321	(151) P3$_1$12	(152) P3$_1$21
		(153) P3$_2$12	(154) P3$_2$21	(155) R32	
	3m	(156) P3m1	(157) P31m	(158) P3c1	(159) P31c
		(160) R3m	(161) R3c		
	$\bar{3}\,m$	(162) P$\bar{3}$1m	(163) P$\bar{3}$1c	(164) P$\bar{3}\,m$1	(165) P$\bar{3}\,c$1
		(166) R$\bar{3}\,m$	(167) R$\bar{3}\,c$		
Hexagonal	6	(168) P6	(169) P6$_1$	(170) P6$_5$	(171) P6$_2$
		(172) P6$_4$	(173) P6$_3$		
	$\bar{6}$	(174) P$\bar{6}$			
	6/m	(175) P6/m	(176) P6$_3$/m		
	622	(177) P622	(178) P6$_1$22	(179) P6$_5$22	(180) P6$_2$22
		(181) P6$_4$22	(182) P6$_3$22		
	6mm	(183) P6mm	(184) P6cc	(185) P6$_3$$cm$	(186) P6$_3$$mc$
	$\bar{6}\,m$2	(187) P$\bar{6}\,m$2	(188) P$\bar{6}\,c$2	(189) P$\bar{6}$2m	(190) P$\bar{6}$2c
	6/mmm	(191) P6/mmm	(192) P6/mcc	(193) P6$_3$/mcm	(194) P6$_3$/mmc
Cubic	23	(195) P23	(196) F23	(197) I23	(198) P2$_1$3
		(199) I2$_1$3			
	$m\bar{3}$	(200) P$m\bar{3}$	(201) P$n\bar{3}$	(202) F$m\bar{3}$	(203) F$d\bar{3}$
		(204) I$m\bar{3}$	(205) P$a\bar{3}$	(206) I$a\bar{3}$	
	432	(207) P432	(208) P4$_2$32	(209) F432	(210) F4$_1$32
		(211) I432	(212) P4$_3$32	(213) P4$_1$32	(214) I4$_1$32
	$\bar{4}$3m	(215) P$\bar{4}$3m	(216) F$\bar{4}$3m	(217) I$\bar{4}$3m	(218) P$\bar{4}$3n
		(219) F$\bar{4}$3c	(220) I$\bar{4}$3d		
	$m\bar{3}\,m$	(221) P$m\bar{3}\,m$	(222) P$n\bar{3}\,n$	(223) P$m\bar{3}\,n$	(224) P$n\bar{3}\,m$
		(225) F$m\bar{3}\,n$	(226) F$m\bar{3}\,c$	(227) F$d\bar{3}\,m$	(228) F$d\bar{3}\,c$
		(229) I$m\bar{3}\,m$	(230) I$a\bar{3}\,d$		

* The number in parenthesis preceding each space group symbol represents their order of listing in International Tables of X-ray Crystallography – Vol. 1.

** The Symbols for the monoclinic system shown here are of the second setting where the 2-fold axis coincides with the b-axis. In the first setting, the 2-fold axis coincides with the c-axis. Accordingly, symbols starting with C will start with B; e.g., B2, Bm, etc.

Table 3.5   Schoenflies Notation for Space Groups

No.	Sch.	No.	Sch.	No.	Sch.	No.	Sch.	No.	Sch.
(1)	C_1^1	(47)	D_{2h}^1	(93)	D_4^5	(139)	D_{4h}^{17}	(185)	C_{6v}^3
(2)	C_i^1	(48)	D_{2h}^2	(94)	D_4^6	(140)	D_{4h}^{18}	(186)	C_{6v}^4
(3)	C_2^1	(49)	D_{2h}^3	(95)	D_4^7	(141)	D_{4h}^{19}	(187)	D_{3h}^1
(4)	C_2^2	(50)	D_{2h}^4	(96)	D_4^8	(142)	D_{4h}^{20}	(188)	D_{3h}^2
(5)	C_2^3	(51)	D_{2h}^5	(97)	D_4^9	(143)	C_3^1	(189)	D_{3h}^3
(6)	C_s^1	(52)	D_{2h}^6	(98)	D_4^{10}	(144)	C_3^2	(190)	D_{3h}^4
(7)	C_s^2	(53)	D_{2h}^7	(99)	C_{4v}^1	(145)	C_3^3	(191)	D_{6h}^1
(8)	C_s^3	(54)	D_{2h}^8	(100)	C_{4v}^2	(146)	C_3^4	(192)	D_{6h}^2
(9)	C_s^4	(55)	D_{2h}^9	(101)	C_{4v}^3	(147)	C_{3i}^1	(193)	D_{6h}^3
(10)	C_{2h}^1	(56)	D_{2h}^{10}	(102)	C_{4v}^4	(148)	C_{3i}^2	(194)	D_{6h}^4
(11)	C_{2h}^2	(57)	D_{2h}^{11}	(103)	C_{4v}^5	(149)	D_3^1	(195)	T^1
(12)	C_{2h}^3	(58)	D_{2h}^{12}	(104)	C_{4v}^6	(150)	D_3^2	(196)	T^2
(13)	C_{2h}^4	(59)	D_{2h}^{13}	(105)	C_{4v}^7	(151)	D_3^3	(197)	T^3
(14)	C_{2h}^5	(60)	D_{2h}^{14}	(106)	C_{4v}^8	(152)	D_3^4	(198)	T^4
(15)	C_{2h}^6	(61)	D_{2h}^{15}	(107)	C_{4v}^9	(153)	D_3^5	(199)	T^5
(16)	D_2^1	(62)	D_{2h}^{16}	(108)	C_{4v}^{10}	(154)	D_3^6	(200)	T_h^1
(17)	D_2^2	(63)	D_{2h}^{17}	(109)	C_{4v}^{11}	(155)	D_3^7	(201)	T_h^2
(18)	D_2^3	(64)	D_{2h}^{18}	(110)	C_{4v}^{12}	(156)	C_{3v}^1	(202)	T_h^3
(19)	D_2^4	(65)	D_{2h}^{19}	(111)	D_{2d}^1	(157)	C_{3v}^2	(203)	T_h^4
(20)	D_2^5	(66)	D_{2h}^{20}	(112)	D_{2d}^2	(158)	C_{3v}^3	(204)	T_h^5
(21)	D_2^6	(67)	D_{2h}^{21}	(113)	D_{2d}^3	(159)	C_{3v}^4	(205)	T_h^6
(22)	D_2^7	(68)	D_{2h}^{22}	(114)	D_{2d}^4	(160)	C_{3v}^5	(206)	T_h^7
(23)	D_2^8	(69)	D_{2h}^{23}	(115)	D_{2d}^5	(161)	C_{3v}^6	(207)	O^1
(24)	D_2^9	(70)	D_{2h}^{24}	(116)	D_{2d}^6	(162)	D_{3d}^1	(208)	O^2
(25)	C_{2v}^1	(71)	D_{2h}^{25}	(117)	D_{2d}^7	(163)	D_{3d}^2	(209)	O^3
(26)	C_{2v}^2	(72)	D_{2h}^{26}	(118)	D_{2d}^8	(164)	D_{3d}^3	(210)	O^4
(27)	C_{2v}^3	(73)	D_{2h}^{27}	(119)	D_{2d}^9	(165)	D_{3d}^4	(211)	O^5
(28)	C_{2v}^4	(74)	D_{2h}^{28}	(120)	D_{2d}^{10}	(166)	D_{3d}^5	(212)	O^6
(29)	C_{2v}^5	(75)	C_4^1	(121)	D_{2d}^{11}	(167)	D_{3d}^6	(213)	O^7
(30)	C_{2v}^6	(76)	C_4^2	(122)	D_{2d}^{12}	(168)	C_6^1	(214)	O^8
(31)	C_{2v}^7	(77)	C_4^3	(123)	D_{4h}^1	(169)	C_6^2	(215)	T_d^1
(32)	C_{2v}^8	(78)	C_4^4	(124)	D_{4h}^2	(170)	C_6^3	(216)	T_d^2
(33)	C_{2v}^9	(79)	C_4^5	(125)	D_{4h}^3	(171)	C_6^4	(217)	T_d^3
(34)	C_{2v}^{10}	(80)	C_4^6	(126)	D_{4h}^4	(172)	C_6^5	(218)	T_d^4
(35)	C_{2v}^{11}	(81)	S_4^1	(127)	D_{4h}^5	(173)	C_6^6	(219)	T_d^5
(36)	C_{2v}^{12}	(82)	S_4^2	(128)	D_{4h}^6	(174)	C_{3h}^1	(220)	T_d^6
(37)	C_{2v}^{13}	(83)	C_{4h}^1	(129)	D_{4h}^7	(175)	C_{6h}^1	(221)	O_h^1
(38)	C_{2v}^{14}	(84)	C_{4h}^2	(130)	D_{4h}^8	(176)	C_{6h}^2	(222)	O_h^2
(39)	C_{2v}^{15}	(85)	C_{4h}^3	(131)	D_{4h}^9	(177)	D_6^1	(223)	O_h^3
(40)	C_{2v}^{16}	(86)	C_{4h}^4	(132)	D_{4h}^{10}	(178)	D_6^2	(224)	O_h^4
(41)	C_{2v}^{17}	(87)	C_{4h}^5	(133)	D_{4h}^{11}	(179)	D_6^3	(225)	O_h^5
(42)	C_{2v}^{18}	(88)	C_{4h}^6	(134)	D_{4h}^{12}	(180)	D_6^4	(226)	O_h^6
(43)	C_{2v}^{19}	(89)	D_4^1	(135)	D_{4h}^{13}	(181)	D_6^5	(227)	O_h^7
(44)	C_{2v}^{20}	(90)	D_4^2	(136)	D_{4h}^{14}	(182)	D_6^6	(228)	O_h^8
(45)	C_{2v}^{21}	(91)	D_4^3	(137)	D_{4h}^{15}	(183)	C_{6v}^1	(229)	O_h^9
(46)	C_{2v}^{22}	(92)	D_4^4	(138)	D_{4h}^{16}	(184)	C_{6v}^2	(230)	O_h^{10}

* Space group number as in Table 3.4

# Structural Principles in Crystals

4.1 INTRODUCTION

The previous chapters have been devoted to the symmetry in crystals which are mostly based on the geometrical concepts. Although mention was made about the building blocks, no rationale has been presented as to how the crystals are built out of atoms or molecules. In this chapter, we will focus our attention on the details about the structural principles of the building up of crystals. Undoubtedly, such a presentation should also be able to explain the external symmetry in terms of the internal structures.

We know that matter exists in three states—gas, liquid and solid. There are atoms or molecules in all the three states of matter. However, the way in which they are held together differs in each state. In the gaseous state, the molecules are separated by large distances which depend on the pressure. For example, at one atmosphere, they are approximately 30 Å apart, on a statistical basis. Consequently, the attractive or repulsive interactions between the molecules are negligible. The molecules in the gaseous state are therefore free to move in any direction and will completely occupy the available space. In the liquid state, the molecules are only a few angstroms apart and the interactions between the molecules are stronger than in a gas. As a result, a liquid takes the shape of the vessel it is contained in, but it need not fill the available space completely. Around a given molecule in a liquid there are a few neighbouring molecules; when we look around another molecule there may not be an identical arrangement. Such orderliness is limited to small regions in a liquid and is called *short-range order*. In a solid, the distances between the molecules are nearly the same as in a liquid, but the interaction between the molecules is much stronger. This restricts the movement of molecules to a limited region—the molecules cannot move past the neighbouring molecules. The arrangement of molecules around any one of them is identical to that around any other molecule, even when such molecules are situated far apart. Such an arrangement is called *long-range order*. Many distinguishable properties of crystals are a result of this long-range order. The molecules mentioned above can be made up of one atom (He, Ar, Ni, Zn, Cu), or two (N_2, O_2, CO), or more than two (Na_2O, CO_2) atoms. Molecules may also be replaced by ions, i.e., atoms or molecules with positive or negative charges, e.g. Na^+, Cl^-, CO_3^{2-}, PO_4^{3-}.

Under the structural principles in crystals there are two aspects to be discussed: (1) the geometry of molecular arrangements, (2) the thermodynamic factors that decide such arrangements. In this chapter we will be concentrating on the first aspect, whereas the second aspect is related to the nature of the chemical bonds involved. The latter, in turn, is dependent on the electronic arrangements in the atoms. There are three types of chemical bonds, namely: (1) when the electrons are transferred from one type of atoms to another, the resulting ions can hold together because of the electrostatic attraction. This results in the ionic bond, e.g. a NaCl crystal is formed from Na^+ and Cl^- ions. (2) If the outer electrons of the two atoms are shared in between, instead of the transference; this bonding is called a covalent bond. Such a sharing of electrons results in a definite geometrical configuration of the atomic arrangement in rigid structures, e.g. covalently bonded carbon atoms in diamond or graphite crystals. (3) The outer electrons of a large number of atoms (in the crystal) belong to the crystal as a whole rather than sharing between atoms. At the same time, the positively charged nuclei collectively retain the electrons within the crystal. This type of bonding, called metallic bonding, involves strong interactions. There can be much weaker molecular interactions in crystals, such as those with hydrogen bonding or van der Waal's bonding. The latter leads to the molecular crystals which are characteristically soft and have low melting points, e.g. hydrogen bonding in ice crystals or van der Waal's bonding in wax crystals.

The detailed discussion on various aspects of chemical bonding, although highly essential for a student of crystallography, cannot be dealt with in detail in this treatise. Hence, the reader may refer to the recommended books listed at the end.

4.2   GEOMETRY OF MOLECULAR ARRANGEMENT

4.2.1   Close Packing

When identical atoms or molecules in a crystal are represented as hard spheres of equal size, they can be arranged in three dimensions in different ways. Those arrangements which fill the available space most efficiently are called close packing. The arrangement should be such that one sphere is in actual contact with the maximum number of similar spheres. In fact, close packing can be demonstrated in many objects around us, such as a bunch of grapes, a pile of oranges, a honeycomb, a raft of soap bubbles, etc.

There is no space between the hexagonal cells in a honeycomb. In close packing, we can think of spherical atoms being arranged such that we have a rigid solid with minimum space between the atoms. The arrangement of spheres can be studied by considering the spheres in one plane. Fig. 4.1a has each sphere in contact with four neighbours. The space occupied by an individual sphere can be represented by a cube (Fig. 4.1b).

The volume of the sphere $= \frac{4}{3}\pi r^3$ (4.1)

The volume available for each sphere $= (2r)^3 = 8r^3$ (Fig. 4.1b)

$$\text{Percentage of space filled by the sphere} = \frac{\text{Vol. of sphere}}{\text{Vol. of the cube}} \times 100 \quad (4.2)$$

$$= \frac{(4/3)\pi r^3}{8r^3} \times 100 = 52.4\%$$

In comparison, consider the arrangement shown in Fig. 4.1c. In one layer, each sphere is in contact with six spheres, and each sphere is surrounded by six depressions or voids. Each void, in turn, is surrounded by three spheres. The space occupied by a sphere can be represented by a hexagonal prism (Fig. 4.1d).

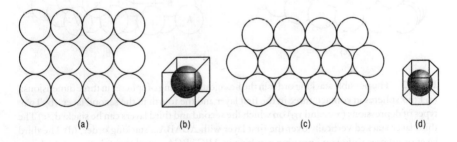

(a)　　　　(b)　　　　(c)　　　　(d)

Fig. 4.1 Close packed arrangement of spheres: (a) Each sphere touching four spheres, (b) The space needed for a sphere, defined by a cube, (c) Second type of planar arrangement of spheres with each sphere touching six spheres, (d) The space needed for the sphere, defined by a hexagonal prism

$$\text{Percentage of space filled by the sphere} = \frac{\text{Vol. of sphere}}{\text{Vol. of the hexagonal prism}} \times 100$$

$$= \frac{(4/3)\pi r^3}{\left(2\sqrt{3}\,r^2\right)2r} \times 100 = 60.5\%$$

Therefore, the arrangement in Fig. 4.1d has a higher percentage of filling and a larger number of nearest neighbours for each sphere. In fact, the hexagonal packing is the closest packing of spheres. The number of neighbours nearest to each sphere is 12 in three dimensions. The piling up of close packed spheres shown in Fig. 4.1c can be achieved in more than one way. In order to understand this, let us consider four spheres as in Fig. 4.2a.

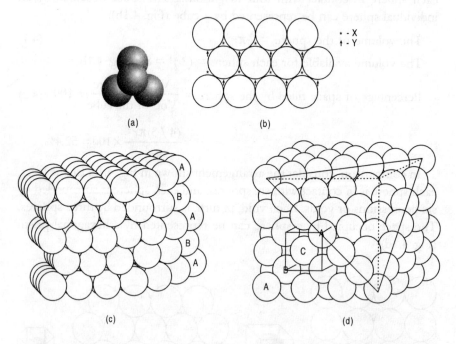

Fig. 4.2   The possible stacking order in the hexagonal packing of spheres in three dimensions: (a) Four spheres in contact, three in the first layer and the fourth in the second layer, (b) Two types of depressions (*xxx* and *yyy*) on which the second and third layers can be stacked, (c) The third layer stacked vertically over the first layer with ABABA... stacking order, (d) The third layer of spheres sitting on *y* position resulting in ABCABCA... stacking order are of the layers parallel to (111) is shown together with a unit cell of F-cubic lattice

If we keep three of the spheres touching each other, there is a depression formed—the fourth sphere can be accommodated over it. The three spheres belong to the first layer, whereas the fourth belongs to the second layer. By placing the fourth sphere in the depression, not only is a stable arrangement produced but also all the spheres are touching each other. Now, let us extend this arrangement introducing more spheres and layers. A layer of close packed spheres sits over another layer with the same packed arrangement. Here, we have two choices for the placement. The centres of the spheres belonging to the second layer may lie either over the depressions marked *xxx* or over those marked *yyy* (Fig. 4.2b). The spheres of the third layer can be added in two alternative positions. Either they can be vertically over the spheres of the first layer marked A (Fig. 4.2c), or, they can be vertically over the *yyy* positions (Fig. 4.2d).

Let us designate the first layer as A, the second layer whose spheres are occupying the *xxx* position as B, and the layer with the spheres sitting over the position *yyy* as C. For the arrangement in Fig. 4.4c, the stacking order of

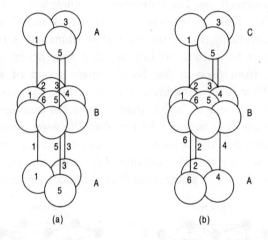

(a)   (b)

Fig. 4.3   Three dimensional view of (a) ABABAB... stacking in HCP and (b) ABCABC... in CCP. The numbers refer to the depressions formed between the spheres.

the layers will be ABABABABA... and the stacking order in Fig. 4.2d will be ABCABCABCABCA... The above close packed arrangement will be clearer in the front view with the minimum number of spheres. Thus, in Fig. 4.3, there are seven spheres in the middle layer, arranged in a close packing. Around the central sphere, there are six depressions which are numbered 1 to 6. Three spheres are placed on the depressions 1, 3, 5 from the top. Three other spheres can be placed from the bottom in two ways. Either they can go to 1, 3, 5 from the bottom (Fig. 4.3a), or, they can take the positions 2, 4 and 6, again from the bottom (Fig. 4.3b). It is evident from Fig. 4.3 that the total number of nearest spheres remains the same (i.e. twelve) in either of the arrangements. This is also called the coordination number. In the first case, the spheres added on the top as well as the bottom are identical in their orientation; these layers can be designated as *A*, and the middle layers can be designated as *B*, so that the stacking sequence is ABABABABA... In the second case, the two added layers are not identical in their orientation and have to be designated separately as, say A and C. Thus, the arrangement in the second case will be ABCABCABCA... We shall see in the next section, the ABABA... stacking will be a hexagonal close packing (HCP) and the ABCABCA... stacking will form a cubic close packing (CCP).

4.2.2 Symmetry of Close Packing

The two types of close packing, stacking arrangements, discussed in section 4.2.1, differ significantly in symmetry characteristics. The arrangement in Fig. 4.2a has the symmetry $3m$. The symmetry of a single layer about a sphere is $6mm$ (Fig. 4.2b). This holds good for layers A, B and C when they are considered individually, but when the AB combination is considered the spheres will not be vertical to one another and, hence, the symmetry decreases, i.e. from $6mm$ to $3m$. So the space group of AB (two-layer formation) is P3m. If the combination is ABA (three layers taken together), then the symmetry will be higher than P3m as the number of symmetry elements increases, but they will be less than P6/mmm (hexagonal Bravais lattice). In reality, the ABA arrangement does not possess a true 6-fold axis but only the 6_3 screw axis. This is because of the presence of the extra spheres in the middle of the cell of a hexagonal Bravais lattice (compare Fig. 4.4a with

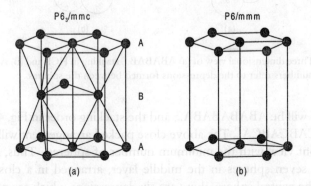

Fig. 4.4 Comparisons of ABA arrangement with hexagonal-centred Bravais Lattice: (a) shows the 6_3 screw axis instead of 6-fold axis but it is a HCP lattice (b) has a 6-fold axis but is not a HCP lattice

the hexagonal Bravais lattice, Fig. 4.4b). Similarly, all the original mirror planes of external morphology are only glide planes to the c-axis (c-glide). Therefore, the space group of the ABA arrangement is P6$_3$/mmc. This being a hexagonal space group, the arrangement ABABA is called hexagonal close packing (HCP). It can be stated that a hexagonal Bravais lattice is not a HCP lattice.

In order to understand the ABCABCA type of stacking, we should reorient the layers in a different way—the planes containing the layers are tilted by 45° from the horizontal (Fig. 4.2d). Consider four layers ABCA (Fig. 4.5) in the tilted position. Remove all the spheres, except one, from the A-layers at the top and bottom; six spheres each from B and C layers. This will enable us to outline a cube. The relation between the orientation of the cube and the individual layers can be seen from Fig. 4.5. The 3-fold axis of the cube therefore coincides with the c-axis of the hexagonal packing. Each face of the cube has an additional sphere, thus delineating a unit cell of a face-centred cube lattice. All the symmetry elements of the space group, $Fm3m$, are present in this unit cell. Therefore, the stacking of ABCA... is called cubic close packing (CCP). Thus, we see that the space groups of the most common close packings are $P6_3/mmc$ (HCP) and $Fm3m$ (CCP). We may wonder whether there are other space groups for which the close packing is possible. The answer is that there are 8 space groups for close packing:

$Fm3m$	$R\bar{3}m$
$P6_3/mmc$	$P\bar{3}m$
$P6_3mc$	$R3m$
$P3/mm2$	$P3m.$

As you can see, there is only one close packing which belongs to the cubic system while the rest belong to the hexagonal and trigonal systems; of these

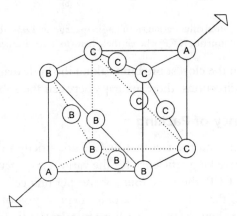

Fig. 4.5   The ABCA... arrangement tilted 45° from the horizontal and outlined by a cube. The 3-fold axis of the cube coincides with the 6-fold axis of HCP. ABCACA... is a CCP ($Fm3m$).

the HCP (ABABA) has already been discussed. The other space groups correspond to the different types of stacking sequences such as ABAC..., etc. Details of these stacking sequences are not elaborated in this monograph, as they are rarely encountered in common crystal structures.

4.2.3   Body-centred Cubic Packing

Besides the close packing discussed in the previous section, there can be other types of packing. The best known example encountered in many crystals is the body-centred cubic packing (BCC). The arrangement is shown in Fig. 4.6a. Consider piling up of the layers shown in Fig. 4.1a in such a way that the spheres of the second layer sit in the depression formed between the four spheres of the first layer. The third layer will be a repetition of the first layer. This leads to a cubic unit cell, with one sphere at the body centre and the other at the corners of the cube, resulting in the space group symmetry I*m3m*. The number of nearest spheres (coordination number) in this type of arrangement is 8 as compared to 12 in the case of HCP and CCP lattices. It is evident, therefore, that BCC is not a close packed arrangement. This is also obvious from Fig. 4.6a, where the spheres do not touch in all the directions. In fact,

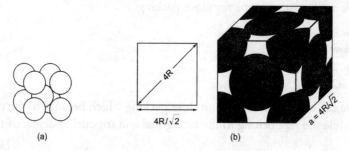

(a)   (b)

Fig. 4.6   (a) Body-centred cubic packing resulting in space group I *m3m* (b) Position of spheres in a cubic close packed structure (CCP) wherein the spheres touch each other on all face diagonals.

the spheres are in the closest proximity along the body diagonal of the cube. In all the other directions, there are gaps between the spheres.

4.2.4   Efficiency of Packing

In order to understand the fact that BCC is not strictly a close packing, we can calculate the packing efficiency for the various arrangements of spheres.

In the case of CCP, there are four spheres per unit cell. The *packing factor* corresponding to that of the CCP arrangement = Vol. of spheres/Vol. of the unit cell. The volume of the unit cell can be calculated in terms of the radius of the sphere as shown in Fig. 4.6b. Here, the length of the side of a cube = $4R / \sqrt{2}$.

$$\text{Packing factor} = \frac{4\left(\frac{4}{3}\pi R^3\right)}{\left(4R / \sqrt{2}\right)^3} = 0.74 \qquad (4.4)$$

In other words, 74% of the available space is filled by a CCP arrangement.

In the HCP arrangement, the unit cell contains 6 spheres. Here, only 1/3 of the hexagonal unit cell is considered in terms of the rhombic prism shown in Fig. 4.7a. This effectively contains two spheres. Of these two, one sphere is in the midpoint of the rhombic prism. In order to determine the cell volume in terms of the radius of the spheres, we have to calculate the cell edges in terms of the radius of the spheres.

The horizontal cell edge $a = 2R = $ The length of the side of the tetrahedron. The vertical cell edge c is equal to twice the height of the tetrahedron shown in Fig. 4.7a.

The vertical height of the tetrahedron $= 1.633R$

The vertical cell edge $c = 2(1.633R)$

The volume of the rhombic trigonal prism $= 2\left(\sqrt{3}R^2 h\right)$

$$= 2\sqrt{3}R^2 \times 2(1.633R)$$

Packing factor $=$ Vol. of spheres/Vol. of rhombic prism

$$= \frac{2\left(\frac{4}{3}\pi R^3\right)}{2\sqrt{3}R^2\left(2(1.633R)\right)} = 0.74 \tag{4.5}$$

Thus, the packing factors are the same for both CCP and HCP.

In the case of BCC, there are two spheres in the unit cell.

The edge of the cube $a = 4R/\sqrt{3}$ as shown in Fig. 4.7b.

$$\text{The packing factor} = 2\left(\frac{4}{3}\pi R^3\right)\Big/\left(4R/\sqrt{3}\right)^3 = 0.68 \tag{4.6}$$

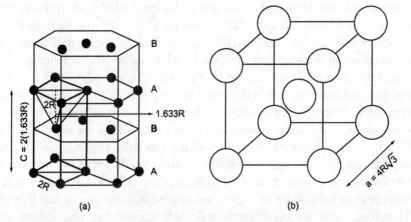

Fig. 4.7 (a) The HCP (ABABA...) arrangement considered, wherein a rhombic trigonal prism is delineated as the unit cell (b) BCC arrangement

Table 4.1   The parameters, such as packing efficiency, coordination number and the shortest interatomic distances encountered in the different types of packing of spheres

Arrangement	Shortest atomic distance	Packing efficiency	Coordination number
Cubic close packing (CCP)	$a_0\sqrt{2}/4$	0.74	12
Hexagonal close packing (HCP)	$a_0/\sqrt{2}$	0.74	12
Body-centred close packing (BCC)	$a_0\sqrt{3}/4$	0.68	8
Simple cube	$a_0/\sqrt{2}$	0.52	6

Thus the BCC structure is not a close packing. The packing factor goes down to 0.52 if we consider a simple cubic lattice. Table 4.1 shows the various parameters encountered in commonly known packings of spheres.

4.2.5   Voids within Close Packing

As shown in Table 4.1, even the most effective packing leaves nearly 26% of empty space. This is because the spheres are considered as hard spheres and cannot be deformed or squeezed to fill these voids. Understanding the distribution and relative sizes of the voids is important in the case of crystals containing two or more elements (atoms of different sizes). Let us consider the four spheres shown in Fig. 4.8a If the centres of the spheres are joined by straight lines, we get a tetrahedron which surrounds the void enclosed by the four spheres. Hence, the void is called a tetrahedral void. Another type of void that can be recognized in a close packing is the one surrounded by six spheres. When the centres of these six spheres are connected, it results in an octahedron surrounding this void and is called the octahedral void (Fig. 4.8b). A careful examination of Figs. 4.8, 4.9a, and 4.3 reveals that only two kinds of voids present in close packings are of tetrahedral and octahedral symmetry. In the close packing of spheres we have seen that there are six depressions surrounding a given sphere on either side of the layer (Fig. 4.3). When a second layer of spheres is added from the top, three of the six depressions become tetrahedral voids. Similarly, on the lower side three more tetrahedral voids are generated. In addition, the central sphere will form two more tetrahedral voids with the second nearest layer on the top and bottom. Therefore, there are eight tetrahedral voids surrounding one sphere, and each tetrahedral void is surrounded by four spheres.

Ratio of tetrahedral voids/spheres = 2   (4.7)

Similarly, the three alternate depressions around one sphere become octahedral voids on the top. Correspondingly, there will be three more such voids on the lower side. Thus, we have six octahedral voids surrounding each sphere, and six spheres around each void.

Ratio of octahedral voids/spheres = 1 (4.8)

It can be quickly deduced that there are eight tetrahedral and four octahedral voids in a unit cell of a CCP lattice because there are four spheres in a unit cell. Likewise, there are twelve tetrahedral and six octahedral voids in a HCP unit cell which contains six spheres. Thus, in both the cases the ratio of tetrahedral voids: octahedral voids: number of spheres per unit cell is equal to 2 : 1 : 1.

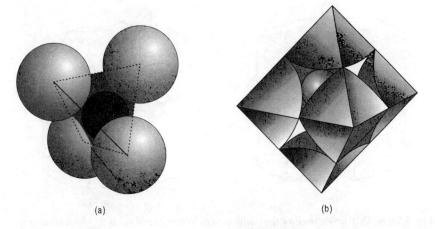

(a) (b)

Fig. 4.8 (a) The tetrahedral void at the centre of four spheres (b) Octahedral void at the centre of six spheres. The octahedron is formed by joining the centres of six spheres

The location of these voids in a CCP lattice is shown in Fig. 4.9a. The octahedral voids are located at the middle of the cell edges and at the centre of the cubic unit cell. Since tetrahedral voids are formed by connecting the centres of the nearest four spheres, there are four tetrahedral voids in the upper half and four in the lower half of the cubic cell.

Fig. 4.9b shows the location of octahedral voids in the HCP unit cell. It is clear from this figure that the octahedra are inclined with respect to the *c*-axis, and one of the apices of each of the octahedron needs the sphere from the adjoining unit cell. All the octahedra share faces with one another and the octahedra of the adjoining cells share spheres from the first cell. On an average, the number of octahedra per unit cell remains six, i.e. three in the upper half and three in the lower half of the unit cell. Six tetrahedral voids can easily be outlined inside the hexagonal unit cell (Fig. 4.10c). There are two tetrahedral voids along the vertical lines that result when we join the three

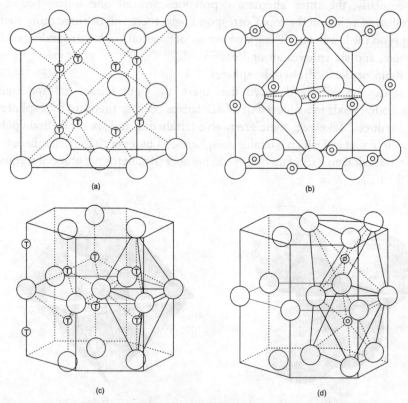

Fig. 4.9   (a) CCP lattice to show the position of eight tetrahedral voids (b) Octahedral voids in CCP unit cell (c) Tetrahedral voids in HCP unit cell (only two sites on the edges are shown) (d) Octahedral voids in HCP unit cell (only two are shown)

central spheres with the spheres on the centres of the upper and lower faces. Besides, there are twelve tetrahedral voids situated on the edges of the hexagon. Each of these tetrahedral voids is shared by three of the adjoining unit cells. Therefore, the total number of tetrahedral voids belonging to a hexagonal unit cell is 6 + 2 + (12/3) = 12.

4.2.6   The Concept of Radius Ratio

Besides the symmetry and location of the voids, we are also interested in their sizes, as the voids will be occupied by atoms of second constituents. The radius ratio is a relative number—it is the ratio of radius of the sphere that can occupy the void to that of the sphere which goes to build the close packing.

Besides tetrahedral and octahedral voids, there can be other types of voids with different coordinations. For example, there can be two, three, seven eight, or twelve coordination. In the 2-fold linear coordination (Fig. 4.10a), the centres of the two spheres lie on a straight line.

The radius of the larger sphere is designated as R_X and that of the smaller one as R_A. The ratio R_A/R_X should be small in the case of a close packing. In fact, two-fold coordination is non-existent in close packing. The three-fold coordination produces a triangular void (Fig. 4.10b), wherein the centres of all the three spheres are in a plane.

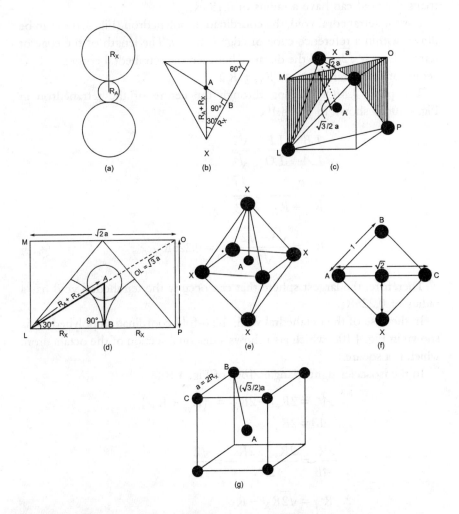

Fig. 4.10 Illustration of the coordination number and its relation to the size of the void: (a) 2-fold linear coordination (b) a 3-fold coordination producing a triangular void (c) a tetrahedral void (d) diagonal plane of a reference cube used to calculate the radius ratio of tetrahedral void (e) an octahedral void (f) a diagonal section in octahedron which is a square (g) an 8-fold coordination

From Fig. 4.10b,

$$XB/AX = R_X / (R_X + R_A) = \cos 30°$$

$$R_A = R_X / (\cos 30 - R_X) = 0.155 R_X$$

or $R_A / R_X = 0.155$ $\qquad(4.9)$

This means that the largest sphere that can be accommodated in a triangular void can have a radius of $0.155 R_X$.

For the tetrahedral void, the coordination polyhedron (Fig. 4.10c) can be drawn within a reference cube of edge length a. The length of the edge of tetrahedron $= \sqrt{2}a$ and the distance between the tetrahedral corner and the centre of the tetrahedron $= (R_X + R_A)$

The plane $LMOP$ passing through the centre of the tetrahedron in Fig. 4.10c is shown in Fig. 4.10d.

$$\frac{LB}{LA} = \frac{LP}{LO} = \frac{\sqrt{2}}{\sqrt{3}}$$

$$\frac{R_X}{R_A + R_X} = \frac{\sqrt{2}}{\sqrt{3}}$$

$$R_A = \frac{\sqrt{3} - \sqrt{2}}{\sqrt{2}} R_X = 0.225 R_X \qquad(4.10)$$

Therefore, the largest sphere that can occupy the tetrahedral void has a radius of $0.225 R_X$.

In the case of the octahedral void, the 6-fold coordination polyhedron is shown in Fig. 4.10e, which also shows a diagonal section of the octahedron, which is a square.

In the isosceles right triangle ABC, of Fig. 4.10f

$$AC = 2R_A + 2R_X = 2(R_A + R_X)$$

$$AB = 2R_X$$

$$\frac{AC}{AB} = \frac{2R_A + 2R_X}{2R_X} = \frac{\sqrt{2}}{1}$$

$$\therefore R_A = \sqrt{2}R_X - R_X$$

$$= 0.414 R_X \qquad(4.11)$$

The maximum radius of the spherical occupant in the octahedral void is $0.414 R_X$.

Although a primitive cube is not a close packing, we may consider the 8-fold coordination by placing an atom in the centre of the cube (Fig. 4.10g). Here, $AB = R_A + R_X$, and the cell edge $= 2R_X$ which is equal to a.

$$BC = 2R_X = a \text{ or } R_X = a/2$$

$$\text{Body diagonal} = 2AB = (2\sqrt{3}/2)a = a\sqrt{3}$$

$$AB = a\sqrt{3}/2$$

$$R_A + R_X = a\sqrt{3}/2, R_A = a\sqrt{3}/2 - R_X = a\sqrt{3}/2 - a/2 = a/2(\sqrt{3}-1)$$

$$\therefore R_A = R_X(\sqrt{3}-1) = 0.732\,R_X \tag{4.12}$$

The maximum radius of the sphere that can occupy the void with 8-fold coordination is $0.732R_X$.

It can be shown that 12-fold coordination can be achieved only when $R_A = R_X$ (See Figs. 4.3a-b and Fig. 4.6b).

It can be seen from the above that the size of the void increases with the coordination number. For a given coordination number there is a limiting range for the size of the sphere occupying the void (R_A). This is summarized in Table 4.2.

Table 4.2 The relationship between the size of the void (expressed as radius ratio) and the coordination number

Description	Radius ratio	Coordination numbers Maximum X ions Surrounding A ions
Cubic close packing and Hexagonal close packing	1.00–0.73	12–8
Six corners of an octahedron	0.73–0.414	6–4
Four corners of a tetrahedron	0.414–0.225	4
Corners of triangle	0.225–0.155	3
Linear on opposite sides of A	< 0.155	2

The size of the voids mentioned in Table 4.2 is not valid for other types of packing. As an example, let us consider the BCC arrangement. There are two types of voids in this lattice, one of these occurs at the positions shown in Fig. 4.11a, wherein two adjacent cubes are necessary for completing the tetrahedron. There will be four such voids around each sphere, and around each void there are four spheres. Therefore, the ratio of voids to spheres is 1 : 1. However, it is easy to see that the tetrahedron drawn in Fig. 4.11a is distorted. This is because two edges of each of the triangular sides have lengths 2R, while the third one is a (= length of the cube edge). It can be shown that,

$$R_A = 0.291R_X \qquad (4.13)$$

Similarly, consider the octahedral voids in the BCC structure shown in Fig. 4.11b. Here again, the two adjacent cubes define the octahedron, i.e. taking two central atoms and four corner atoms. This, again, is a distorted octahedron as is evident from the figure. It can be shown that

$$R_A = 0.154R_X \qquad (4.14)$$

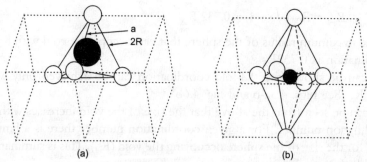

(a)   (b)

Fig. 4.11 Voids formed in two adjacent BCC lattices: (a) distorted tetrahedron where one of the edge is equal to *a*-parameter and other edges equal to 2R (b) a distorted octahedron formed by the sharing of four spheres on the interface of adjacent cubes and two from the body centre

We can see that the radius ratio of the octahedral site in a BCC structure is smaller than that of the tetrahedral site. Since there are six such voids around each sphere at the face centres not shown in Fig. 4.11b and an octahedral void is surrounded by six spheres, the void to sphere ratio is 1 : 1. Therefore, the ratio of the number of tetrahedral voids to the number of octahedral voids to the number of spheres is 2 : 1 : 1 in BCC structure. This is the same as in the HCP and CCP structures.

4.3 SIMPLE TYPE STRUCTURES

The principles of close packing discussed above can help in understanding the crystal structures of some crystalline solids. They are called "type structures" because they serve as the basis from which more complex structures can be derived. Although there are a large number of type structures, only the simplest are chosen here for discussion with the aim of introducing the subject to beginners. The type structures are classified on the basis of the ratio of the constituents (*A* and *X*) of the crystals. For example, AX, AX_2, A_2X_3, AX_3, etc.

4.3.1 AX Structures

Under this type we will discuss the structures of sphalerite and wurtzite (both being polymorphs of ZnS), NaCl, CsCl, and niccolite (NiAs).

4.3.1.1 Sphalerite (Zinc blende) Structure (F$\bar{4}$3m)

This structure is shown in Fig. 4.12a. Here, the sulphur atoms form the CCP lattice and the alternate tetrahedral voids are filled with Zn. Thus, each Zn atom is surrounded by four sulphur atoms and, in turn, each sulphur atom is surrounded by four Zn atoms. There are four sulphur atoms in a unit cell

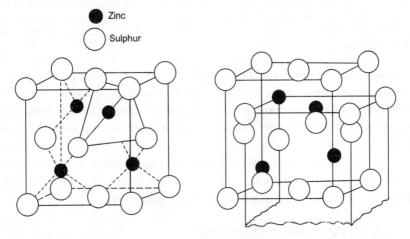

Fig. 4.12 (a) Sphalerite (ZnS) structure (F$\bar{4}$3m). Zn fills alternate tetrahedral voids of CCP lattice of S (b) Sphalerite structure shown as two penetrating CCP lattices of Zn and S, known as sublattices

together with four Zn atoms. The Zn : S ratio is therefore 1 : 1 which is in agreement with the composition, ZnS. In fact, it can be shown that the Zn atoms in a sphalerite structure separately form a CCP lattice, if we connect the Zn atoms from the adjoining unit cell (Fig. 4.12b). Therefore, the sphalerite structure can be called an interpenetration of two CCP lattices. Since both these lattices are parts of the sphalerite structure, they are called sublattices, i.e. sphalerite consists of a Zn sublattice with a CCP arrangement and a S sublattice, again with CCP arrangement.

If both sublattices of the sphalerite structure are occupied by the same kind of atoms, the resulting structure will be comparable to that of a diamond. The difference is that all the tetrahedral positions are occupied by carbon atoms, i.e. $R_A/R_X = R_A/R_A = 1$ (here, X and A have no difference). Indeed, this

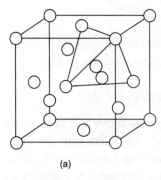
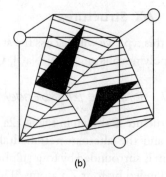

(a)   (b)

Fig. 4.13 (a) The derivation of a diamond structure from sphalerite, when both Zn and S are replaced by C (b) The four tetrahedra in a diamond structure. Only three out of six spheres at corners are shown because they belong to other tetrahedra. The symmetry reduces from $F\bar{4}3m$ to $Fd3m$

structure can be considered to be made up of corner sharing tetrahedra, wherein each carbon atom is surrounded by four carbon atoms (Fig. 4.13 a-b). In Fig. 4.13 b, the four atoms shown at the corners do not form part of the four tetrahedra but are related to those in the other tetrahedra because of the so called diamond glide. Hence, the space group of the distorted lattice is $Fd3m$ as against $F\bar{4}3m$ of sphalerite. A large number of elements or compounds adopt the diamond structure, of which the well-known ones are silicon and germanium (the elemental semiconductors).

4.3.1.2   Wurtzite Structure (P6₃mc)

This is a polymorphic form of ZnS (compounds having different crystal structures but the same chemical composition). The wurtzite structure consists of hexagonally close packed sulphur atoms, wherein the alternate

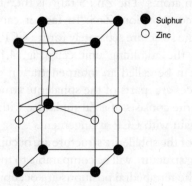

● Sulphur
○ Zinc

Fig. 4.14 Wurtzite Zn structure ($P6_3mc$) showing one-third of a unit cell. Each Zn atom is surrounded by four S atoms

tetrahedral voids are filled by the Zn atoms. Each sulphur atom is surrounded by four Zn atoms, and, in turn, each Zn atom is surrounded by four sulphur atoms (Fig. 4.14). Thus, wurtzite can be shown to consist of two interpenetrating HCP sublattices of S and Zn atoms.

The difference between the wurtzite and sphalerite structure is in the localized arrangement. This can be shown to arise from the difference in the orientation of the neighbouring tetrahedra. In wurtzite, the corners of two adjacent tetrahedra are oriented in the same direction, while in sphalerite they are oriented in opposite directions.

A large number of compounds possess the same crystal structure as that of sphalerite (CuCl, CuBr, ZnS, CdSe, CdS, AlSb, GaP, etc.). Such crystals are said to be isostructural. Similarly, compounds such as BeO, ZnO, CdS, AlN, MgTe, GaN, InSb, etc. are isostructural with wurtzite.

4.3.1.3 Halite (NaCl) Structure (Fm3m)

The halite structure can be regarded as a cubic close packing of Cl^- ions. All the octahedral sites are filled by Na^+ ions as shown in Fig. 4.15. Each Cl^- ion is surrounded by six Na^+ ions, and each Na^+ ion is surrounded by six Cl^- ions to produce a 6 : 6 coordination. Na^+ and Cl^- constitute the two interpenetrating

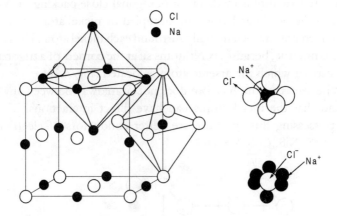

Fig. 4.15   Halite (NaCl) structure (F$m3m$) with 6-fold coordination of Na^+ and Cl^-

CCP sublattices of Cl^- and Na^+ ions. The alternate planes will be made up of one kind of atoms, i.e. either Na^+ or Cl^-. $R_{Na}/R_{Cl} = 0.54$ in the case of halite, and this value falls within the range of 0.41–0.73 which is the radius ratio of a stable octahedral coordination (see Table 4.2). A large number of compounds adopt the NaCl structure, which include oxides, selenides, sulphides, halides, etc. Although most of these compounds possess a radius ratio within the prescribed limits, there are some exceptions, such as LiI (0.35), BaO (0.96) or CsF (1.26). The reason for these exceptions is the appreciable covalent rather than ionic character of the bonding involved.

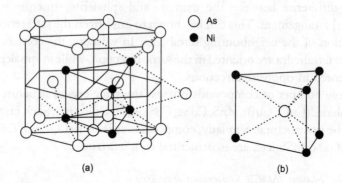

(a)   (b)

Fig. 4.16   (a) Niccolite (NiAs) HCP structure (P6₃/*mmc*) where six As atoms octahedrally surround the Ni atoms (b) The Ni atoms centred on the corners of a trigonal prism

4.3.1.4 Niccolite (NiAs) Structure (P6₃/mmc)

The niccolite structure is made up of hexagonal close packing of As atoms in which all the octahedral voids are occupied by nickel atoms (Fig. 4.16a). Here the arsenic atoms octahedrally surround each nickel atom. However, the converse is not true, because six Ni atoms sit at the corners of a trigonal prism at the centre of which the arsenic atom is located (Fig. 4.16b).

It can be seen that the As atom does not sit midway between any two Ni atoms and that the metal atoms can have direct interactions. Therefore, crystals possessing this structure are better conductors of electricity. For example, FeS, NiS, CoS, VS, etc.

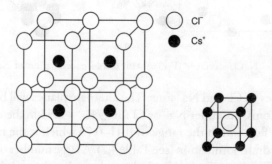

Fig. 4.17   (a) Four unit cells of CsCl structure (P*m*3*m*) where each Cs⁺ ion is in 8-fold coordination with Cl⁻ (b) Cl⁻ surrounded by eight Cs⁺

4.3.1.5 The Structure of CsCl (Pm3m)

This is made up of a primitive cubic lattice of Cl^- ions, the centre of the lattice being occupied by the Cs^+ ion. Fig. 4.17 shows eight Cl^- ions surrounding each Cs^+ ion and eight Cs^+ ions surrounding each Cl^- ion, thus having a 8:8 coordination. $R_{Cs^+}/R_{Cl^-} = 0.92$. The solids which adopt CsCl structure will usually have a radius ratio > 0.73, e.g. CsI (0.76), NH_4Cl (0.76), etc. The CsCl lattice is, therefore, not a close packed structure.

4.3.2 AX$_2$ Structures

Here, we will consider the structures of $CdCl_2$, CaF_2, Cu_2O, and TiO_2.

4.3.2.1 CdCl$_2$ Structure $(R\bar{3}m)$

This crystal contains a cubic close packed structure of chloride ions. The smaller cations occupy only 50% of the octahedral sites. As in the case of the NaCl structure, the octahedral voids are the layers on the (111) plane. In the

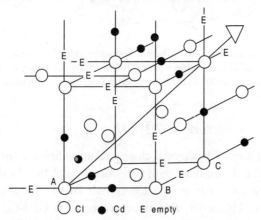

○ Cl ● Cd E empty

Fig. 4.18 The $CdCl_2$ structure $(R\bar{3}m)$ where Cd occupies half of the octahedral sites. In the CCP array of Cl^- anions, 'E' indicates empty octahedral sites. Only one trigonal axis exists unlike in the NaCl structure where four 3 axes exist, hence it is a trigonal system.

$CdCl_2$ structure, alternate planes corresponding to the octahedral voids are completely empty. Therefore, the structure consists of one Cd layer sandwiched between two adjoining chloride layers (Fig. 4.18). Because the next cation layer corresponding to the octahedral voids is missing, the two chloride layers will face each other. Since they are the same type of ions they repel each other, and are held together only by weak van der Waals forces. The crystal can, therefore, be easily cleaved along this layer. Another consequence of this repulsion is the distortion of the cubic symmetry resulting in a rhombohedral symmetry. Many metal hydroxides of composition $M(OH)_2$ adopt this layered type structure; so do the disulphides such as ZrS_2, SnS_2, TiS_2, etc. In metal hydroxides, the repulsion is between the OH layers. In the sulphides, the repulsion between the S layers causes easy cleaving. Another consequence is the strong anisotropic thermal expansion.

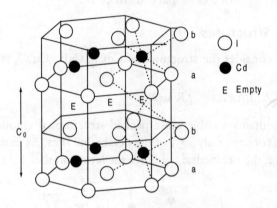

Fig. 4.19   The Cdl_2 structure ($P\overline{3}m1$), where Cd occupies half of the octahedral sites in the HCP array of anions. 'E' indicates empty sites. The first 50% occupancy eliminates sharing of faces between octahedra unlike in niccolite where Ni occupies all the octahedral sites and octahedra share faces

4.3.2.2 Cdl_2 Structure ($P\overline{3}m1$)

This structure consists of hexagonal close packed iodine ions in which 50% of the octahedral voids are filled with Cd^{2+} ions (Fig. 4.19). As shown in the figure, the octahedral voids in the lower half of the hexagonal cell are completely empty. Therefore, this structure consists of I-Cd-I sandwich layers. Here again, the two iodine layers are facing one another, hence the ions are held together only by weak van der Waal's forces. This results in similar properties—cleaving, etc. as in the case of $CdCl_2$.

4.3.2.3 CaF_2 Structure ($Fm3m$)

This structure can be described in two alternate ways. In the first way, we can say that the Ca^{2+} ions form a cubic close-packed arrangement, with all the eight tetrahedral voids in the cubic cell being filled by fluoride ions. Since there are four calcium ions belonging to the unit cell, the ratio of Ca : F is 4 : 8 or 1 : 2 (Fig. 4.20a). In the second way (Fig. 4.20b), the structure can be described in terms of a simple cubic lattice of fluoride ions with the cations

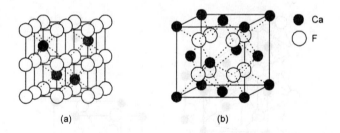

Fig. 4.20 Structure of fluorite, CaF_2 (*Fm3m*): (a) F⁻ ions are in simple cubic packing with Ca in the centre of a cube (b) Standard unit cell of fluorite

occupying the cube centre. Therefore, there are eight fluoride ions surrounding each Ca ion, while only four Ca ions surround a given F⁻ ion. The fluorite structure is adopted by AX_2 compounds, where R_A/R_X normally exceeds 0.73, although there are exceptions. For example, BaF_2 (1.01), SrF_2 (0.84), $BaCl_2$ (0.74) and ThO_2 (0.73) are normal fluorite structures, whereas HfO_2 (0.56), ZrO_2 (0.56) and CeO_2 (0.67) are exceptional in possessing the fluorite structure.

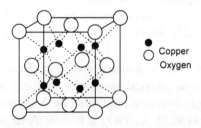

Fig. 4.21 Structure of Cu_2O (*Fm3m*). Note the exchange of the cations position compared to the fluorite structure in Fig. 4.20 b, and hence known as the antifluorite structure

4.3.2.4 Cu_2O *Structure (Fm3m) [Cuprite]*

In the cuprite structure, the oxygen ions are arranged in a cube where all the tetrahedral voids are filled by the cations, i.e., Cu⁺ (Fig. 4.21). The ratio of cation : anion ($A : X$) is 2 : 1, and this is therefore considered as an example of A_2X type structure. Other examples are Na_2O, Li_2S, Li_2O, etc. Comparing this arrangement with that of CaF_2, it is possible to understand that Cu⁺ ions form a simple cubic lattice in which the oxygen atoms are located at the cube corners. As a result of this, the cuprite structure is called an antifluorite structure. Therefore, such a reversal of positions of anions and cations in the structure leads to an antistructure or antitype. This is a common feature encountered in many other crystal structures.

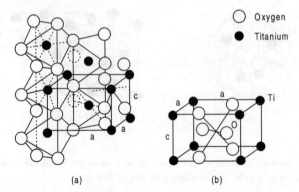

(a)   (b)

Fig. 4.22 Structure of rutile TiO_2 ($P4_2/mnm$): (a) The edge-sharing octahedra parallel to the c-axis and corner-sharing along *a* and *b* crystal direction, (b) The rutile structure adapted by AX_2 structures with cations directly facing each other

4.3.2.5 TiO_2 Structure $(P4_2/mnm)$ [Rutile]

This structure contains octahedrally coordinated cations, i.e. Ti^{4+} ions, surrounded by six oxygen ions, with each oxygen ion surrounded by three titanium ions. Thus, it differs considerably from the NaCl structure. Therefore, the symmetry is changed to the tetragonal system. The octahedra share the corners in the a- and b- directions (Fig 4.22a), whereas in the c-direction they share the edges.

The metal ions in the c-direction can have direct interaction (Fig. 4.22b) giving rise to special physical properties in compounds having the rutile structure. The rutile structure is adopted by AX_2 compounds, where R_A/R_X is in the range 0.41–0.73. In TiO_2, it is 0.49. A large number of solids are isostructural to rutile, for example SnO_2, MnO_2, MgF_2, PbO_2, etc.

The structural description of solids with other types such as A_2X_3, AX_3 or ABX_3, AB_2X_4 is avoided in this introductory treatise for reasons of brevity and the interested reader may refer to the recommended books.

4.4 PAULING'S RULE

Most crystal structures discussed so far can be adopted by minerals and solid inorganic compounds. Such structures are largely adopted by ionic compounds. The principles behind this description can be summarized by what are known as Pauling's rules.

4.4.1 First Rule

The radius ratio of A and X determines the nature of the coordination polyhedron and the coordination number (see Table 4.2).

4.4.2 Second Rule

The ionic crystal is stable only if the sum of the charges on the anions equals the sum of the charges on the cations.

In the NaCl structure, each Na^+ ion is surrounded by six Cl^- ions. Therefore, the unit positive charge is divided into 1/6th part towards each Cl^- ion. Similarly, each Cl^- anion is surrounded by six Na^+ ions, therefore the unit negative charge of Cl^- is divided into 1/6 parts toward each Na^+. In effect, the total charge will be $+(1/6) \times 6 = -(1/6) \times 6$. This is valid irrespective of the complexities of the composition or structure. The consequence of this rule is that the crystal as a whole should be neutral. This principle of electroneutrality is violated in the case of different solids where the charge neutrality is attained by additional electrons or holes (positive charge).

4.4.3 Third Rule

The sharing of edges, more particularly the faces, by two anionic polyhedra decreases the stability of ionic crystals. This is explained through the Fig. 4.23. The distance between the cations decreases as one moves from corner-sharing to edge-sharing to face-sharing, thus increasing the covalent interaction between the cations. This is exemplified in the silicate structures, where Si^{4+} ions are always separated from one another by an oxide ion, thereby retaining the ionic character in silicates. The tendency is greater when the charges on

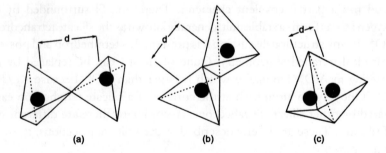

(a)	(b)	(c)

Fig. 4.23 The cationic distance (d) between neighbouring MO_4^+ tetrahedra decrease from corner-sharing (a) to edge-sharing (b) to face-sharing (c). Hence ionic bond strength is a>b >c

the cations are larger, for example Si^{4+}, Nb^{5+}, W^{6+}, Mo^{6+}, etc. Incidentally, these ions adopt smaller coordination numbers.

4.4.4. Fourth Rule

In a crystal structure containing different cations, those of high valency and small coordination number will not share polyhedron elements such as edges and faces with each other. The fourth rule is an extension of the third and the justifications of the latter are valid in this case as well.

4.4.5   Fifth rule

The number of essential constituents of a crystal are always limited to a minimum. In other words, the occupancy of the voids in a periodically packed structure is a minimum even if the voids are left vacant; solids prefer to leave them so rather than being filled by neutral atoms, even if the size permits.

4.5   SILICATE STRUCTURES

Silicates are compounds containing Si^{4+} and O^{2-} and are the major constituents of the earth's crust. Therefore, the study of silicate structures is important to students of mineralogy. The treatment in this book cannot be exhaustive because silicate structures are fairly complex. Therefore, only an introductory treatment of the topic is given in this section. The aim is merely to show that they can be discussed under structural principles.

All silicate structures consist of oxygen ions. Some of them adopt close packing while others are less closely packed. The voids in the oxygen packing are filled by suitable cations, of which Si is the most abundant. The radius ratio R_{Si}/R_O is 0.28, favouring tetrahedral coordination. Besides, the Si–O bond has a partial covalent character. Therefore, Si surrounded by four oxygen atoms forms a stable unit, generally known as the silicate tetrahedron— SiO_4^{4-}. In any structural discussion it is presented as tetrahedron and possesses near-ideal tetrahedral geometry. Quite often Si^{4+} can be replaced by Al^{3+}, resulting in AlO_4^{5-} tetrahedra which is larger than SiO_4^{4-} because $R_{Al}/R_O = 0.35$. One oxygen atom each at the corners of a silicate tetrahedron can be shared by other silicate tetrahedrons. Accordingly, there are different types of silicates. These are briefly described in the following sections.

4.5.1   Nesosilicates (Island Silicates)

Here, none of the corners of the tetrahedron are shared, i.e. the SiO_4^{4-} tetrahedrons remain isolated from one another by cations other than Si (Fig. 4.24a). They are called nesosilicates or isosilicates. In principle, the nesosilicates are salts of the hypothetical acid, H_4SiO_4. The well-known examples are the olivine group of minerals (A_2SiO_4) and the garnets $[A_3B_2(SiO_4)_3]$.

4.5.2 Sorosilicates (Double-island Silicates)

Here, two SiO_4^{4-} tetrahedra share one oxygen atom so that the composition corresponds to $Si_2O_7^{6-}$ (Fig. 4.24b). The oxygen atom shared between two SiO_4 tetrahedra is called the bridging oxygen atom, whereas those at the unshared corners are non-bridging oxygen atoms. The sorosilicates, also called double-island silicates, are relatively rare. For example, hemimorphite $(Zn_4Si_2O_7(OH)_2)$ and thortveitite $(Y_2Si_2O_7.H_2O)$ and epidote $X_2Y_3O(SiO_4)(Si_2O_7)(OH)$.

4.5.3 Cyclosilicates (Ring Silicates)

Here three, four, or six member rings of the general formula Si_xO_{3x} can form ring silicates (Fig. 4.24c). The three-member ring is shown by the mineral benitoit $(BaTiSi_3O_9)$. The four-member ring is observed in axinite $(Ca_3Al_2BO_3Si_4O_{12}.OH)$ and the six-member ring in beryl $(Be_3Al_2Si_6O_{18})$. The cyclosilicates are also called ring silicates.

4.5.4 Single-chain Silicates

Here, each SiO_4^{4-} tetrahedron shares two corners with the adjoining tetrahedra to form a chain (Fig. 4.24d). These chains run parallel to each other in the crystal structure and have the effective composition SiO_3^{2-}. The single-chain silicate structures are exemplified by the pyroxenes $ASiO_3$ or $AB(SiO_3)_2$. A and B are the cations which link two or more parallel chains. The single-chain silicates are also called meta-silicates.

4.5.5 Double-chain Silicates

When the two-chain silicates shown in Fig. 4.24e share the corners in such a way that there are two types of tetrahedra, those sharing corners with two neighbouring tetrahedra (marked X) and those which share corners with three neighbouring tetrahedra (marked Y). Thus, the effective composition corresponds to $Si_4O_{11}^{6-}$. Minerals of the amphibole groups are the best known examples of double-chain silicates. Hence, the double chains are linked by other cations as well as anions $(OH)^-$, F, etc. Therefore, we have amphiboles such as anthophyllite $(Mg_7Si_8O_{22}(OH_2)$ or tremolite $(Ca_2Mg_5Si_8O_{22}(OH)_2)$.

Both single- and double-chain silicates are also called inosilicates (*ino* means thread).

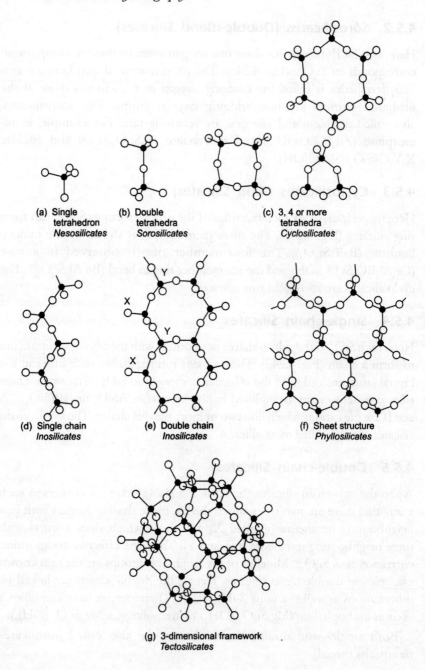

(a) Single
tetrahedron
Nesosilicates

(b) Double
tetrahedra
Sorosilicates

(c) 3, 4 or more
tetrahedra
Cyclosilicates

(d) Single chain
Inosilicates

(e) Double chain
Inosilicates

(f) Sheet structure
Phyllosilicates

(g) 3-dimensional framework
Tectosilicates

Fig. 4.24　Silicate structures formed by the sharing of SiO_4 tetrahedra

4.5.6 Layer Silicates (Sheet Silicates)

Each SiO_4^{4-} tetrahedron is sharing 3 corner oxygens, whereas one corner is unshared. The effective composition is $Si_2O_5^{4-}$ or $(Si_4O_{10}^{4-})$. This type of silicate sharing (Fig. 4.24f) leads to sheet-like structures exemplified by talc $(Mg_3Si_4O_{10}(OH)_2)$, pyrophillite $(Al_2Si_4O_{10}(OH)_2)$, muscovite $(KAl_2AlSi_3O_{10}(OH)_2)$, etc. The hydroxyl ions in layer silicates form a part of the layered structure and can be replaced by F^-, Cl^-, etc. The cations Mg, K, Al bridge one silicate structure to another. The lengths between the sheets are weak compared to the Si-O bond. Hence, they are easily cleaved parallel to sheets and are called phyllosilicates (*phyllo* means leaf).

4.5.7 Framework Silicates (Tectosilicates)

Here, all the four corners of the SiO_4^{4-} tetrahedra are shared with the four neighbouring tetrahedra (Fig. 4.24g). If there is no substitution for Si, then the structure has the composition SiO_2, which takes a 3-dimensional network often termed as the framework silicates (quartz, tridymite, crystabolite). If 25% of the Si^{4+} are substituted by Al^{3+}, the framework has to be charge-neutralized with monovalent cations, e.g. albite $(NaAlSi_3O_8)$ or orthoclase $(KAlSi_3O_8)$. If 50% of Si^{4+} is replaced by Al^{3+}, then the framework substitution requires a divalent cation, e.g. anorthite $(CaAl_2Si_2O_8)$.

4.6 DEFECTS IN CRYSTALS

The discussion so far has indicated that crystal structures are perfect in their 3-dimensional atomic arrangements. However, it is well known that physical systems including gases and liquids deviate from ideal behaviour. Solids are no exceptions to this generalization.

The deviation from ideality in crystal structure will be localized in most cases. Such deviations, in general, are called imperfections or defects. Structural defects can be classified into:

1. Point defects—isolated absence or misplacement of atoms
2. Line defects—misplacement along an array or line
3. Plane defects—misplacement or absence along the plane
4. Place exchange defects—ordered/disordered arrangement of atoms.

In addition to these, the effects of thermal fluctuations on atoms and the presence of electrons or positive holes are considered as defects. Even the presence of an extra atom on the surface of a crystal is a defect. Details of these lattice types of defects will be dealt with in any textbook of solid state chemistry or physics.

4.6.1   Point Defects

Let us consider the crystals of metals and alloys where the atoms are neutral. If the atoms are missing from the regular lattice sites, it is called a vacancy. In this case, it is assumed that the missing atom has been taken away from the solid. It is also possible that the misplaced atom is located at a void (interstice) adjoining the vacancy. In ionic solids, such a defect is called the Frenkel defect (Fig. 4.25c).

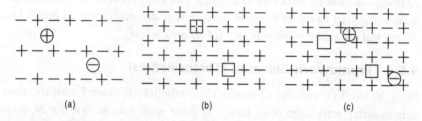

Fig. 4.25   Point defects: (a) Interstitial defect (b) Schottky defect (c) Frenkel defect

Extra atoms can also be present without a vacancy in the neighbourhood; this is called an interstitial defect. Usually such atoms are other than those of the lattice, i.e. impurities of an atomic size different from that of the lattice atoms (Fig. 4.25a).

In ionic crystals lattice atoms are charged. They are made up of anions and cations. It is possible to remove cations and anions from the regular lattice positions in the crystals. Such defects are called Schottky defects (Fig 4.25b).

For the sake of electrical neutrality, the anion and cation vacancies should be equal in number. So also, the interstitial ions in the case of Frenkel defect should be equal to the vacancies. In many a case, this may not be true. Then the charge neutrality is maintained by the introduction of electrons or positive holes. Therefore, the presence of point defects (Frenkel and Schottky defects) modify many of the physical properties.

The Schottky and Frenkel defects can also be produced by impurity atoms, i.e., foreign atoms carrying an electrical charge different from that of the lattice ions. For example, Ca^{2+} substituting for Na^+ in the NaCl structure, or O^{2-} substituting for Cl^- in the same crystal. In the first case, an equal number of Na^+ vacancies have to be produced, and in the second case Cl^- vacancies equal to O^{2-} ions have to be generated for maintaining the charge neutrality. In many of the crystals, point defects are present during their formation so that the chemical composition is somewhat altered. For example, CoO has fewer metal ions than expected, i.e. it may be $Co_{1-x}O$; the mineral maghemite has the composition Fe_2O_{3-x} containing fewer oxygen atoms than expected. This phenomenon is known as *non-stoichiometry*.

4.6.2 Line Defects

These are also called dislocations because they are along a line in the crystal. There are two types of dislocations: (a) Edge dislocation (b) Screw dislocation.

An edge dislocation is formed if a plane of atoms is terminated half-way in the interior (Fig. 4.26a). The planes remaining in the upper half of this diagram are relatively in an expanded state, whereas those in the lower side are under compression. Therefore, the external stress on the location of the extra plane can be shifted; for example, in the case of shearing of metal wires leading to breakage.

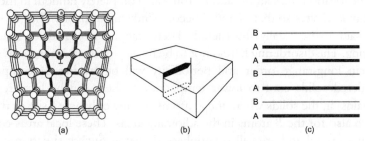

(a)　　　　　　　　(b)　　　　　　　　(c)

Fig. 4.26 Line defects: (a) Edge dislocation (b) Screw dislocation (c) Plane defect—stacking fault

The screw dislocation is a defect produced by the introduction of a distorted polyhedron along a line, as though there is a localized screw axis which is not normally present in the ideal structure. This introduction of a screw axis is local and does not affect the overall space group. This can be alternatively visualized in the following way (Fig. 4.26b): The ideal lattice is cut half-way through and the two cut portions are moved in opposite directions and joined back again, so as to retain the continuity of the lattice. However, the lattice structure is not the same in the immediate vicinity of the edge, (i.e. the half-step). This edge continues to provide favourable sites for further addition of atoms so that in the event of crystal growth, the screw dislocations provide favourable sites for growth. This leads to a growth of spirals, often visible on the surface.

4.6.3 Plane Defects

In the case of close packing, we have discussed the stacking of different layers to obtain different close packings such as *ABABAB*... in the case of HCP, and *ABCABCABC*... in the case of CCP. If the sequence is disrupted, we get a sequence *ABABAAB*... or *ABCABABABC*...; this is called a plane defect or stacking fault (Fig. 4.26c). It can so happen that the stacking fault itself can be periodic over the entire lattice. In such cases the space group as well as the

X-ray diffraction pattern is changed. This phenomenon is called polytypism and is found in compounds such as ZnS, SiC, etc.

4.6.4 Order-disorder Defects

Consider a solid made up of atoms A and B. These two types of atoms should occupy two sites in the lattice namely, I and II. If all the A atoms occupy type I sites and the B atoms occupy type II sites, the crystal is said to be perfectly ordered. However, it is possible that a few A atoms may occupy type II sites and a few B atoms may occupy type I sites. This leads to disorder. It is possible that the location of A atoms and B atoms are completely random in the type I and type II sites so that A and B become indistinguishable. They crystal is then said to be totally disordered. There can be intermediate states of ordering. The disordering here is not to be mistaken for an amorphous state, wherein long-range order and periodicity will be lacking. If completely disordered solids cool to lower temperatures, there can be local regions (domains) in the solids where the A atoms occupy only type I site or type II site; so also for the B atoms in the adjoining areas. These local areas coexist within a crystal and are called antiphase domains. Such a situation can be illustrated with an alloy of gold having the composition Cu_3Au. In the completely ordered state the gold atoms occupy the cube corners, whereas the Cu atoms occupy the face centres. As seen from Fig. 4.27a, the system

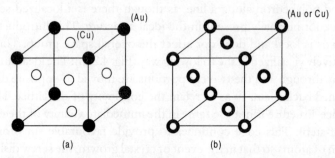

(a)   (b)

Fig. 4.27   (a) An ordered structure of Cu-Au alloy Cu_3Au (*Pm3m*) (b) Disordered structure of Cu-Au alloy where Cu and Au are indistinguishable (*Fm3m*).

corresponding to this ordered arrangement is a primitive cube with a space group P*m3m*. Now imagine that the system is completely disordered and we cannot distinguish between the Cu and Au atoms so that each lattice point can have gold or copper. This leads to the arrangement seen in Fig. 4.27b. This corresponds to a face-centred cube with the space group F*m3m*. This change can be observed by X-ray diffraction. An ordered state shows X-ray reflections in addition to those shown by the disordered state. Therefore, the ordered state is said to form a super lattice.

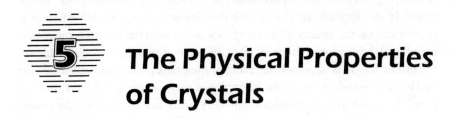

The Physical Properties of Crystals

5.1 INTRODUCTION

All the physical properties of crystals are governed by the nature of the atomic arrangements within the crystal structure and their chemical composition. The physical properties can be directional or non-directional dependent, i.e. they may be anisotropic or isotropic respectively with reference to different crystal directions. For example, density is isotropic, whereas mechanical strength or hardness of crystals can be anisotropic. Elementary knowledge of physical properties such as piezo- or pyroelectricity are essential to determine the presence or absence of inversion symmetry, even if the X-ray diffraction studies have been completed. Optical properties are an integral part of crystallography, because of their direct relation to the symmetry and structure. Besides, the investigation of optical properties is feasible without involving any sophisticated instrumentation. Only certain selected properties are briefly introduced in the following sections. For elaborate treatments on these properties, the reader may refer to the well-known references recommended at the end of this book.

5.2 COHESIVE PROPERTIES

All mechanical properties of crystals come under the cohesive properties. Depending upon the way in which the mechanical force is applied, the crystals react differently by changing their shape so as to absorb this force. Sometimes the change brought into the crystal due to the applied force is reversible (e.g. elasticity).

5.2.1 Elastic Property

The mechanical force applied is generally termed as "stress", expressed as force/unit area. The deformation to this applied force is called "strain" expressed in terms of fraction of percentage of the original crystal dimension (i.e. length, area or volume). If the strain is reversible, it is said to be in the elastic region (Fig. 5.1), i.e. the applied stress is directly proportional to the elastic strain. If the applied stress is above the elastic region, the deformation is permanent, i.e. the change in shape cannot be brought back. This is the plastic region and the change in shape is called the plastic strain. For ductile materials, the plastic region is large before reaching the break point and such solids are said to have high ductility. Some examples of these are metals. A large number of solids do not deform plastically and they break before reaching the plastic region. The ability of a solid to resist plastic deformation is called the yield strength (the maximum value of the load for the beginning of plastic deformation is called the tensile strength). Such solids are non-ductile or brittle materials. All minerals and inorganic solids belong to the group of brittle solids. Their elasticity is an anisotropic property because it depends on the forces of interaction between the constituent atoms and the way in which

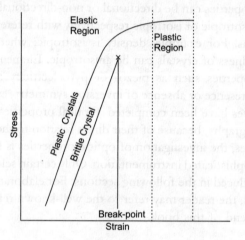

Fig. 5.1   Behaviour of plastic and brittle crystals subjected to stress

the atoms are arranged in different directions within a crystal. The brittleness of solids can be manifested in different ways:

a)   *Fracture* : When a crystal is crushed, it may break into irregular shapes and the fracture surface so produced being: 1) conchoidal, smooth curved surface as in a block of glass, 2) uneven or irregular with rough textured surfaces, 3) hackly, i.e., sharp-edged irregularities such as of metal crystals, 4) splintery, i.e. protrusions similar to broken wood, e.g. asbestos. It is important to note that a fractured surface has no relation to the symmetry, crystal directions or crystal planes.

b)   *Cleavage* : Unlike the fracture, when a crystal breaks along specific crystallographic planes, it is called a cleavage plane. These are invariably parallel to planes of simpler Miller indices, such as basal plane (basal cleavage, 001) or pinacoidal plane (pinacoidal cleavage 100, 110 or 010) (Fig. 5.2).

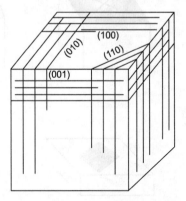

Fig. 5.2 Commonly observed cleavage planes parallel to faces of simple Miller indices, like basal planes, prisms and pinacoids

When cleavage planes are smooth and brilliantly reflecting, they are called perfect cleavages. If there is a step-like combination of two or more cleavages, it is described as an imperfect cleavage.

The property of cleaving is, again, related to crystal structure where the crystal planes with similar type of atoms (made up of only anions or only cations) happen to face one another. Under such conditions interactions between planes are only by van der Waal's forces and the energy required to separate them is very low. The perfect basal plane of mica (001) in the neighbouring layers or the carbon atoms of (0001) planes of graphite repel each other causing perfect cleaving along the basal planes.

c)   *Parting*: A parting plane is also a smooth breaking surface but it differs from cleaving in that, parting occurs along the twin planes. In effect, parting is a type of fracture seen commonly in polysynthetic twinned crystals such as plagioclase. If the crystal is crushed into finer grains, the particles will no more be bounded by smooth surfaces. In contrast, the cleaved surfaces will be observed even after crushing them to fine particles.

d)   *Hardness*: Hardness is a measure of the ability of the solid to withstand aberration or resistance to scratching. The relative hardness is a useful diagnostic property for distinguishing minerals. This means harder minerals scratch the softer ones. Based on this, Mohs (1924) developed a relative scale of hardness involving: Talc (1), Gypsum (2), Calcite (3), Fluorite (4), Apatite (5), Orthoclase (6), Quartz (7), Topaz (8), Corundum (9) and Diamond (10). Diamonds being placed at 10, cannot be scratched by any other minerals and can be used as an abrasive for all other minerals.

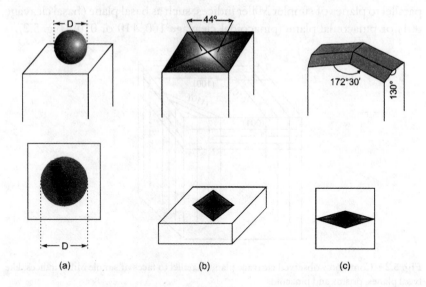

Fig. 5.3 Different shapes of indenters used for testing of hardness on crystal surfaces and the pattern of indentations formed: (a) Brinell's (b) Vicker's (c) Knoop's

Another way to measure hardness is by studying the resistance to indentation under steady pressure. This is called the indentation hardness. The indentation hardness is more quantitative than Mohs' hardness. However, the value will depend upon the shape of the indenter used. Thus, we have: 1) Brinell's hardness—where the indenter is in the form of a hard steel ball; 2) Vicker's hardness—the indenter is in the form of a triangular pyramid and 3) Knoop's hardness—where the indenter is in the form of a elongated square. Figure 5.3 gives the different shapes of indenters. The hardness of crystals is dependent on the existing types of chemical bonding which may by itself differ along the crystal directions. The crystals where van der Waal's forces are predominant are softer than the crystals with ionic bonding. The covalent bonding in three dimensions leads to a high degree of hardness. The hardness also increases with the density of atomic packing in solids. Hardness is also an anisotropic property. Therefore grinding rates vary on the different crystallographic planes of the same crystal.

e)   *Tenacity* : It is a measure of the overall cohesiveness of a crystal—its resistance to breaking, crushing, bending and tearing. It is brittle if it is easily powdered when struck or crushed. It is malleable if it can be flattened to a thin foil or sheet on striking with a hammer. It is sectile if it can be cut into thin shavings and it is ductile if it can be drawn into a wire. It is flexible if it can be bent but does not regain its original size even after the pressure is released.

If it is elastic, it regains the original position upon the release of pressure. A layer of chlorite is flexible while that of mica is more elastic.

5.3 DENSITY AND SPECIFIC GRAVITY

The density of a substance is defined as mass per unit volume, i.e. $\rho = \dfrac{m}{v}$ where m is the mass and v is the volume of a fragment of the substance under consideration at room temperature which is considered to be 20°C. Density can also be calculated by knowing the mass of the unit cell content and the volume of the unit cell of the crystal. The volume of unit cell can be calculated from X-ray diffraction data (discussed in Chapter 6).

The specific gravity of a substance G, is the ratio of the density of the substance at 20°C and the density of water at 4°C. Since the density of water at 4°C and 20°C is very close to unity, for all practical purposes specific gravity and density are not distinguished, except that specific gravity is just a number, while density is expressed in terms of gm/cm^3.

The specific gravity, therefore, is the ratio of mass of a substance to its equal volume of water. It is measured using the Archimedes principle of water displacement, provided it is not soluble in water. In the latter case another appropriate liquid of known density has to be used. Although a common chemical balance can be adopted for determining the specific gravity, a modified Joly's balance or a Berman balance are more convenient for such measurements.

If the crystals are very small, a specific gravity bottle (pycnometer) is used. The methodology for using these techniques is elaborated in most basic books on physics.

The measured density of a substance may sometimes be different from that calculated from X-ray data. This is suggestive of crystal defects, mostly due to point defects in crystals leading to non-stoichiometry. Density is also a diagnostic property related to solid solutions, i.e. in a crystal containing two or more end member compositions within the same crystal structure, e.g. forsterite (Mg_2SiO_4) and fayalite (Fe_2SiO_4) where density increases proportionally with Fe_2SiO_4 content. Pressure and temperature affect the density of a substance. For example, increase in temperature increases the volume without increase in the mass, hence a decrease in density or specific gravity. Increase in pressure decreases the volume without change in mass; hence an increase in density or specific gravity. This has great consequence in understanding the petrology of rocks in the earth's interior, where prevailing high pressures and temperatures considerably affect the densities of minerals and hence their structures. When a given crystal structure is not stable above a particular pressure (or temperatures) it transforms into a new phase with a more stable structure or symmetry. An experimental study of the stability range of such phases as a function of temperature and pressure has led to a better understanding of the conditions of mineral formation in the earth's interior.

5.4   THERMAL PROPERTIES

Those properties which show changes with the supply of the thermal energy come under this category. We are concerned with those properties prevailing under the conditions of low heat energy supply because at higher heat supply, the crystal may melt or decompose.

The specific heat of a solid is an important property and may be defined as the amount of heat required to raise the temperature of a unit mass of a substance by 1°C. This property is again dependent on the type of crystal structure, because the atoms in a crystal have certain vibrational freedom. The heat absorbed from the surroundings enhances these vibrations around the equilibrium positions in the crystal. The vibrations of each atom is not independent on its own, but it is related to the vibrations of other atoms in the same crystal. Hence, it is a collective vibration of the entire lattice. Such lattice vibrations have definite energy values, i.e. they are quantized. The quantized thermal energy value is called a phonon.

Like the electrical energy, heat energy also conducts through the crystals. The thermal conductivity is a quantitative measure of the rate at which the thermal energy is transferred through a crystal which is maintained under a thermal gradient. There are two contributing factors for the thermal conductivity. They are: 1) phonons, i.e. vibrating atoms in a lattice or lattice vibrations, and 2) free electrons. Metals have free electrons and therefore the free electron contribution to thermal conductivity is more in metals.

In non-metals, the thermal conductivity is more for crystalline solids than for amorphous or glassy solids. This is because the collective thermal vibrations are possible only in the crystalline solids. Therefore, glasses are poor thermal conductors as compared to crystalline solids. This property is used by gemmologists to distinguish glass imitations from real gems by just touching the stones with the tip of the tongue. Thermal conductivity is, again, an anisotropic property.

All crystals expand on heating and contract on cooling. This is measured in terms of the thermal coefficient of expansion which is a change in length per unit length (1 cm) of a crystal for a 1° rise in temperature. In isometric crystals, the thermal expansion is the same in all directions, whereas in other systems it is anisotropic. For example, in calcite crystals, the thermal coefficient of expansion is 6×10^{-6} along the a-axis where as it is 25×10^{-6} along the c-axis per degree centigrade. This property can be used to determine the crystallographic directions. Like other physical properties, thermal expansion also depends on the crystal structure. In general, the expansion takes place because of the increase in inter-atomic distances. The crystals with framework structure differ in thermal expansion compared to non-framework structures. Further, the close packed structures show higher thermal expansions compared to those with less close packed structures. Quartz has a lower thermal expansion as compared to periclase (MgO) or copper. Certain solids do not show any thermal expansion particularly in polycrystalline form. This is because the positive expansion in one direction is compensated by the negative expansion in another direction. This is called the thermal compensation effect and such solids do not show any thermal shock effects. Such substances can be quenched fast without development of cracks, e.g. cordierite, tielite, etc. It is a day-to-day experience to see porcelain cracking when it is suddenly cooled from high temperatures, illustrating that such materials have a higher thermal coefficient of expansion.

5.5 ELECTRICAL PROPERTIES

The behaviour of crystals under an applied electric field alone, or combined with thermal energy, is the subject of interest to a very broad area of scientific research. The transport of electrical charges through the crystals gives rise to electrical conductivity, whereas the behaviour of electrical dipoles, particularly in insulators, constitutes the other electrical properties.

5.5.1 Electrical Conductivity

Solids can be classified into metals ($10^6 - 10^8$ ohms^{-1}/m), semiconductors ($10^5 - 10^{-7}$ ohms^{-1}/m) and insulators ($10^{-8} - 10^{-20}$ ohms^{-1}/m), depending upon the range of conductivity exhibited by them. The conductivity of metals decreases with increase in temperature; while the conductivity of semiconductors and insulators increases with temperature. The electrical transport mechanism is mostly electronic and less frequently ionic in nature. In all crystals, the electrical conductivity is an anisotropic property both in electronic as well as ionic conductors. In the case of metals, the free conducting electrons are as many as there are atoms being multiplied by valency. The movement of electrons are restricted by thermal vibrations of the atoms which is greater at higher temperatures. Therefore, the conductivity of metals decreases with an increase in temperature. In semiconductors, the number of the conducting electrons increases with temperature. Thus, in semiconductors although the thermal vibrations of atom exists, the actual number of conducting electrons progressively increases at a tremendous rate with temperature, so that the effective conductivity is larger at higher temperatures.

Both in semiconductors and insulators, the valence electrons have to face an energy gap. If the energy gap is large, then the crystal will be an insulator; if it is smaller, then the crystal behaves like a conductor. This situation can be disturbed by impurity atoms having valencies different from that of the host lattice. Accordingly, we can have intrinsic (without impurities) or extrinsic semiconductors and also insulators with impurity, causing high electrical conductivity. The electrical conductivity of insulators can be enhanced if multivalency of the same element prevails in the system (e.g. Fe_3O_4 containing both Fe^{2+} and Fe^{3+} ions).

The ionic conductivity in crystals is seen mostly in insulators. It arises from the movement of ions instead of electrons. This arises from the point defects namely, Schottky or Frenkels defects, i.e. when the ions move from one vacancy to another or from one interstitial position to another. Ionic conductivity can occur due to large size differences between the cations and anions.

5.5.2   Piezoelectricity

If, in a crystal, one end is not symmetrically related to the directly opposite end (particularly with respect to electrical dipoles), then the crystal is said to possess a polar direction. Eleven crystal classes containing the centre of symmetry do not have a polar axis. On the other hand 21 non-centrosymmetrical classes possess polar directions. Of these 21 crystal classes, all but one (432) possess at least one polar axis with different crystal forms at the opposite end. If mechanical stress is exerted at the ends of polar axis, the electrical charges of opposite sign accumulate at the two ends of the polar directions. This leads to a potential difference developing between the two ends of the same specimen. This phenomenon, known as piezoelectricity, was first discovered by Pierre Curie and Jacques Curie in 1881. The converse of piezoelectricity is possible where, if an AC potential is applied along the polar axis, the crystal will expand along the direction in which the current flows and contracts when the current is reversed. This effect was first detected in quartz and is popularly used even today for electrical oscillators and ultrasound wave detectors. Ultrasound waves are equivalent to the dynamic form of mechanical pressure. Tourmaline is another good example for a piezoelectric crystal which is used as a pressure gauge in shockwave experiments.

5.5.3 Pyroelectricity

A change in temperature in a crystal having a polar axis, produces positive and negative charges at the opposite ends of the polar axis. This phenomenon is called pyroelectricity. Although all the 20 polar classes can exhibit pyroelectricity, some of them are due to the thermal gradient. The thermal stress produced in such crystals can cause a piezoelectric effect and hence the charge generation. Such an effect is called false pyroelectricity. True pyroelectricity is caused by a change of temperature alone and not due to the thermal gradient. This is exhibited only by 10 polar non-centrosymmetry classes which are also called the hemimorphic classes. For example, tourmaline is truly pyroelectric, whereas quartz is not a pyroelectric crystal although it produces the false pyroelectric effect. Therefore all the pyroelectric crystals are piezoelectric, but the converse is not true.

The direction of positive and negative ends of pyroelectric crystals can be reversed by an external electric field. This reversal is a function of strength of the applied electrical field. This phenomenon is called ferroelectricity. This is in complete comparison to ferromagnetic property. Therefore the word ferroelectricity was brought in and the crystals need not contain Fe in any form whatsoever to be ferroelectric. The piezoelectric and pyroelectric properties are vital in detecting the absence of centre of symmetry. Distinguishing between the closely related space groups is done using this property, particularly when it is not possible through X-ray diffraction.

5.6 MAGNETIC PROPERTIES

The behaviour of a crystal to the applied magnetic field can be different depending upon the nature of crystal constituents, i.e., whether they contain transition metal ions, and whether these transition metal ions have strong orientation effects within the crystals. The magnetic lines of forces are repelled away from some crystals. They are called diamagnetic crystals. In other words, the true diamagnetic crystals are repelled out of a magnetic field. A crystal with superconducting property is a true diamagnet. In the second category of crystals, the magnetic lines of force weakly attract them, i.e., the crystal is attracted weakly by the magnet. They are called paramagnetic crystals. With the third category of crystals, the magnetic lines of force are concentrated very strongly in them and they are called ferromagnetic crystals. In this group, the effect of temperature, depending on magnetic attraction, can be different for the various types of crystals. When a ferromagnetic crystal is heated above its critical temperature, it becomes paramagnetic. For example, iron above 770 ° C is paramagnetic and this temperature is called the Curie temperature. In the second type of magnetic materials, the maximum magnetic attraction is noticed at a critical temperature (known as the Neel temperature) and such substances are called antiferromagnetic crystals. Some of the antiferromagnetic crystals are strongly magnetic at lower temperatures and are called ferrimagnets. Most of the iron containing minerals are ferrimagnetic in nature (e.g. magnetite, pyrrhotite, etc.)

All these variations in magnetic properties are related to the spin of the electrons. The electrons behave like micromagnets; however, the property of spin is manifestable only when the electrons are unpaired. That is, they are not occupying the same energy level containing another electron with the opposite spin. The condition of the unpaired nature is normally satisfied only in the case of transition metals and rare earth metal ions. The presence of these type of chemical elements in crystals leads to paramagnetism. Crystals made of non-transitional elements are diamagnetic. The spin on different transitions of rare metal ions can be aligned parallel, giving rise to ferromagnetism. If the spins are aligned exactly opposite, the condition for antiferromagnetism is generated. On the other hand, if the magnitude of the total spin in an antiferromagnetic crystal is not the same, the ferrimagnetism sets in, i.e. one type of spin is larger than the other. This type of spin interaction in all ferromagnetic crystals is anisotropic. Hence, there can be a soft (easy) and a hard direction of the magnetization in the same crystal. In iron, [111] direction is easy and [100] is the hard direction. The differences in magnetic property is made use of in separating minerals in the pure phase from out of their mixtures. The ferrimagnetic properties are prominent as ferrite, which is found in a radio receiver or permanent magnets in loud-speakers.

5.7 OPTICAL PROPERTIES

The interaction of light with crystals manifests a deep complexity unlike in the other two states of the matter—namely liquids and gases. The study of optical properties has its origin in the very early days when man was fascinated with the brilliant colours and varied dispersions of light by precious stones and gems. The systematic understanding of light propagation through crystals led to the development of a strong branch of physics known as Crystal Optics. In fact, the newer discoveries such as that of the solid state LASER has its origin in the principles of crystal optics and this has had a tremendous technological impact on optical communication methods. The optical properties of crystals as a diagnostic tool in distinguishing crystal systems has been practiced for centuries, especially with respect to the identification of crystals under the polarizing microscopes. The nature of light and propagation of light through crystals are, therefore, briefly considered in the following sections.

5.7.1 Nature of Light

Light is a form of energy propagated as an electromagnetic vibration. In order to understand the propagation of light, let us consider the mechanical energy associated with the vibrations of a rope fixed at one end. If the loose end is moved up and down, the waves will progress on the whole length of the rope. These waves traverse in the direction of propagation. In a similar

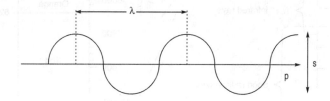

Fig. 5.4 The propagation of light likened to a rope in wave motion, *s*-vibration direction and *p*-propagation direction are mutually perpendicular

way, the light ray has its vibration perpendicular to the direction of propagation. The distance between the two crests (or troughs) of the wave is the wavelength and can be varied in the moving rope by the speed of the movement at the loose end (Fig. 5.4). Similarly, the wavelength of electro-magnetic radiations, denoted by the Greek alphabet λ (lambda), has a very wide range from gamma rays (shortest wavelength) to long radio waves. The wavelength is expressed in nanometers (10^{-9} m) or Angstrom units (Å or 10^{-10} m) or micrometer (10^{-6} m). Figure 5.5 gives the electromagnetic spectrum where the visible part of the spectrum, to which the human eye responds to, has been given in an expanded scale. All the seven colours, which the human eye can distinguish and are contained in the sunlight are shown in the diagram whose width on the scale is proportional to its contribution to form the ray of white light.

The number of vibrations per second is called frequency and is denoted by the alphabet ν (nu). ν is equal to c/λ, where c is the velocity of electromagnetic radiation in vacuum. The frequency of the electromagnetic wave is constant and does not change when it passes through any media.

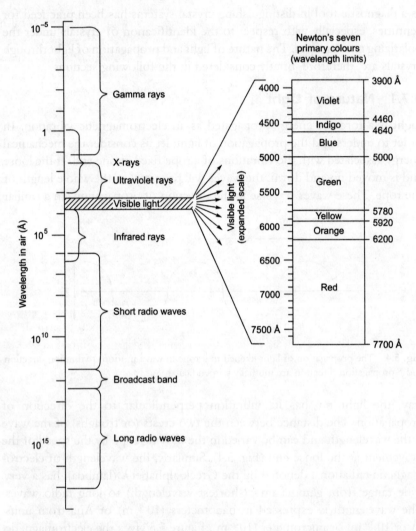

Fig. 5.5   The range of electromagnetic waves. The spectrum of white light has been blown up

However, its velocity is retarded when passing through a medium as compared to a vacuum. In the visible region, the frequency corresponds to about 10^{15} vibrations/sec ($\simeq 10^{15}$ hertz).

5.7.2   Polarization of Light

In Fig. 5.4, which depicts the wave motion on a rope, the vibration is restricted to only one plane. In other words, there is only one plane of vibration which is also the called the vibrational direction. But in electromagnetic waves, there are, in fact, a number of vibration directions which are perpendicular to the line of propagation. This is true, irrespective of whether the light is mono-chromatic or a mixture of various wavelengths. The vibrations of normal light can be restricted to the single plane by special devices called polarizers. This type of light, termed as polarized light, has an analogy with wave propagation on a loose-ended rope but it differs in that the vibration direction in the polarized light corresponds to the electrical vector of the light. Perpendicular to that is the vibration corresponding to the magnetic vector which is also called the polarization direction. In simple terms, an electrical disturbance is always accompanied by a magnetic disturbance perpendicular to it and, thus, together they produce an electromagnetic disturbance. This is what constitutes an electromagnetic wave irrespective of the wavelength. The human eye cannot distinguish between polarized light and natural (unpolarized) light.

5.7.3   Interference

Electromagnetic waves can interfere, just like two or more mechanical waves, under suitable conditions. If two waves satisfy the conditions, namely: 1) same vibration direction (i.e. the same polarization direction), 2) same wavelength and 3) simultaneous origin from the same point of light source, they are said, to be coherent. If the two coherent waves vibrate in-phase, then the resultant amplitude (the magnitude of vibration) adds up to greater intensities (Fig. 5.6a). On the other hand, if the vibrations of the two waves are exactly opposite, i.e. the trough of the first wave coincides with the crest

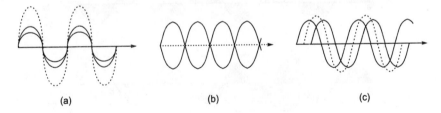

(a)   (b)   (c)

Fig. 5.6   Interferences of light waves of the same amplitude but with (a) phase difference of 1λ (b) phase difference of $\lambda/2$ (c) phase difference of $\lambda/4$.

of the second wave, then both the amplitudes get extinguished (Fig. 5.6b). The phase difference in this is $\lambda/2$, and they are said to be totally out of phase. In the third situation, the two waves can be partly out of phase (i.e., the phase difference is $\lambda/4$ or $3\lambda/4$, etc). The resultant vibration may be greater or smaller than the original amplitude (Fig. 5.6c). If the wavelengths are identical but the vibration directions are different for the two waves, then interference takes place. The resultant light will have an elliptically or circularly polarized wavefront.

5.7.4 Propagation of Light through Crystals

While many crystals transmit light when sliced into thin sections, there are others such as ores, metals, etc. which do not transmit light even in thin sections (opaques). The velocity of light in transparent objects is less than in a vacuum (= 299550 km/s). The ratio of these two velocities is called the refractive index (n) of the substance. For example, the velocity of light in water is 225000 km/sec, so its refractive index (n) is 1.33. Since the refractive index of crystals differ from one another, this property is used for the identification of crystals. The refractive index varies with the wavelength of light as well as temperature. Conventionally, the refractive index of the crystal is measured with sodium light of double wavelengths of 589 and 589.6 nm at 20 ° C, and is denoted as $n^{20°}$.

The propagation of light through a transparent medium results in a phase-displacement which is known as the retardation. In Fig. 5.7, the waves having the same phase pass through media of different refractive index. In the first case, it is passing through the vacuum and there is no retardation. In the second case, it is passing through a medium of definite refractive index (say, $n = 1.5$). The wave has been retarded by $\lambda/2$. Likewise, in the third case, the

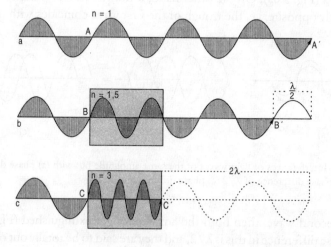

Fig. 5.7 The effect of refractive index of the media on the wavelength of light. The retradation (phase displacement) in: (a) a vaccum is zero (b) in a medium with R.I. = 1.5 is $\lambda/2$ (c) in a medium of R.I. = 3 is 2λ

wave passes through a medium whose $n = 3$. Here, the retardation is by 2λ. In all these cases the wave reached the points A' or B' or C' after these variations. These points are shifted more and more to the left with increasing refractive index. Such retardation is called the phase difference and is measured in terms of the units of wavelengths $(\lambda/2, \lambda/3$, etc.) which are in metric units, i.e. nm. With a decrease in the velocity of the propagation of light (increase in the refractive index) the wavelength also decreases within the medium. However, the colour impression remains unchanged because the wavelength is the same after passing through the media.

5.7.5   Optical Indicatrix

5.7.5a   Isotropic Crystals

In some crystals the velocity of light remains the same in all directions, meaning that the refractive index is the same. Such crystals are called isotropic crystals. The crystals of the cubic system are optically isotropic. For example, garnet ($n = 1.8$), rocksalt ($n = 1.54$), etc. The isotropic crystals are comparable optically to glass, in that both have same refractive index in all directions. A spatial representation of the refractive index in an isotropic crystal would give a sphere (Fig. 5.8), wherein the radii of the sphere is proportional to the

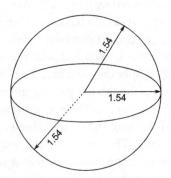

Fig. 5.8   A spatial representation of the refractive index in isotropic substances.

refractive index of the crystals. In other words, infinite vectors radiating from centre of crystal in all directions are proportional in length to the crystals refractive index for light vibrating parallel to the vector direction. The surface connecting the tips of these vectors makes a sphere—also known as "indicatrix".

5.7.5b   Anisotropic Crystals

In crystals of systems other than the cubic system, the velocity of light propagation and hence the refractive index varies with the direction, and all such crystals are said to be anisotropic with respect to light propagation. Anisotropic crystals are uniaxial or biaxial. The optical indicatrix of anisotropic crystals will not be a sphere, but an ellipsoid. The ellipsoid for trigonal, tetragonal and hexagonal crystals will be a rotational ellipsoid with one circular section normal to the longest axis of ellipsoid which coincides with the crystallographic direction. The ellipsoid for orthorhombic, monoclinic and triclinic crystals will be a triaxial ellipsoid with two possible circular sections. The directions perpendicular to the circular sections are called the optic axis or the unique directions. As discussed in the next section, these unique directions or directions of optic axis will not show double refraction. In other words, the crystal behaves like an isotrope along the optic axis direction.

The crystals with single optic axis (or single circular section), are called uniaxial crystals and those with two unique directions (two circular sections of triaxial ellipsoid) are called biaxial crystals. (See sections 5.7.6 and 5.7.7.)

5.7.5c   Double Refraction

This effect was first demonstrated with the help of calcite crystal, wherein an object viewed through it appears as if it has been doubled Fig. 5.9a. When the crystal is rotated, one image remains stationary while the other rotates with the crystal. This phenomenon is known as double refraction. This arises from the fact that each light ray is split into two rays; one of them passes through the crystal in a straight line as though it is passing through an isotropic object and is called an ordinary ray (*o*-ray). The second one is laterally displaced, causing the movement of the image when the crystal is rotated. This is called an extraordinary ray (*e*-ray). The ordinary and extraordinary rays have different velocities of propagation, hence different refractive indices. The difference between the refractive index associated with the extraordinary ray (n_e) and the refractive index associated with the ordinary ray (n_o) is $\Delta n = n_e - n_o$, which is characteristic of different substances. Δn is called the "the birefringence". Because of the difference in the velocities of the two rays, a phase difference is produced which causes the double image. The *o*-ray and *e*-ray will vibrate along two planes which are mutually perpendicular.

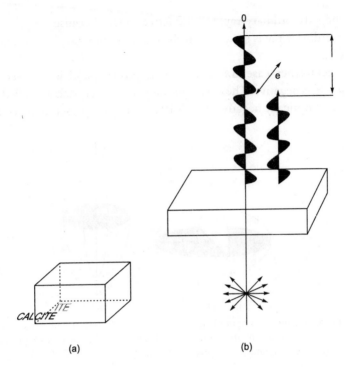

Fig. 5.9 (a) Birefringence or double refraction through a transparent crystal of calcite (b) Schematic representation of the path of light due to double refraction: o-ordinary ray, e-extraordinary ray, the vibration directions of these being mutually perpendicular τ = phase difference.

5.7.6 The Optic Sign in Uniaxial Crystals

As discussed in section 5.7.4, the velocity of light (refractive index) in the various directions of a crystal can be graphically represented by considering the vector lengths originating from the centre of the crystal, which are proportional to the refractive index in that direction. The surface formed by joining all the ends of the vector lengths will define a solid, which is a sphere in isotropic crystals and an ellipsoid in anisotropic crystals. In an anisotropic crystal, because of double refraction, the ordinary ray with equal velocity in all the directions makes a spherical representation of the refractive index, while the extraordinary ray makes a rotational ellipsoid. The velocity of light after double refraction may have two possibilities: 1) The ordinary ray is faster

than the extraordinary ray, i.e. refractive index (because $n \cong \frac{1}{v}$) $n_o < n_e$, 2).
2) The ordinary ray is slower than the extraordinary ray, i.e. refractive index $n_o > n_e$.

In the former case $\Delta n = n_e - n_o$ is positive and in the second case $\Delta n = n_e - n_o$ is negative. The corresponding uniaxial crystals are called optically positive or optically negative. Fig. 5.10 is a planar representation of a section

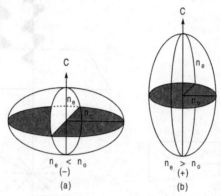

Fig. 5.10   Comparison of optically positive and negative uniaxial indicatrix. x the shaded plane is circular, *C* is the vibration direction of length (which is the c axis/optic axis) (a) for negative crystal $n_e < n_o$ and (b) for positive crystal $n_e > n_o$

of the uniaxial indicatrix. In Fig. 5.10b the n_e is greater than n_o, hence an ellipsoid representing n_e encloses a sphere representing n_o, and hence the crystal is optically positive. In Fig. 5.10a the converse is true, i.e. $n_o > n_e$ and an ellipsoid is enclosed with in the sphere, hence the crystal is optically negative.

5.7.7   Optic Sign in Biaxial Crystals

As discussed in section 5.7.5b, the biaxial indicatrix is an imaginary three dimensional (triaxial) ellipsoid, whose radii are proportional to the refractive indices of crystal for the light vibrating parallel to the directions of radii.

The crystals of orthorhombic, monoclinic and triclinic system belong to the biaxial class. The three mutually perpendicular axis of a triaxial ellipsoid graphically represent the three principal vibration directions of the crystal, and their lengths will be proportional to the maximum, minimum and intermediate values of refractive indices. Conventionally the three principle vibration directions are X representing the minimum refractive index α; Z, representing the maximum refractive index γ; and Y represents the intermediate repractive index β (Fig. 5.11a).

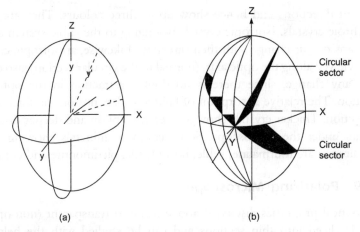

Fig. 5.11   (a) A triaxial ellipsoid showing the variation of the refractive index in a triaxial crystal. The length of radii is proportional to the refractive index. Every vertical plane passing through Z will have one radius equal to β in length, and all these lie in a plane which is circular and normal to the optic axis. There are two such circular sections in a triaxial ellipsoid and hence two optic axis. (b) Distribution of refractive index α, β and γ, in an ellipsoid along the three mutually perpendicular axes X, Y and Z.

The refractive index in any other direction will be between the maximum and the minimum values. The two circular sections in a triaxial ellipsoid intersect along the y-direction (Fig. 5.11b). The directions normal to these circular sections are the unique directions (optic axes) along which no double refraction takes place. The acute angle between two optic axes is called the optic axial angle (2V). The values of 2V can vary between $0°$ to $90°$ and the crystal becomes uniaxial when $2V = 0°$. The two optic axes will lie on the XZ plane and perpendicular to the Y vibration direction or the β-refraction index direction. Hence, Y is called an optic normal. Along this direction, the birefringence $\Delta n = n_\gamma - n_\alpha$ is maximum. If the accute optic axial angle is bisected by n_γ direction or Z-axis, then it is optically positive (+) and when n_α or X direction forms the acute bisectrix, the crystal is optically negative (–).

5.7.8   Pleochroism and Dichroism

Under the polarized light, most of the isotropic crystals exhibit the same absorption for both *o-* and *e*-rays. However, in anisotropic crystals, the absorption depends on the vibration direction as well as the propagation direction. Uniaxial crystals of coloured minerals have the possibility of showing different absorption in two directions leading to a two-coloured appearance; hence, they are called dichroic. Biaxial crystals of coloured minerals have the possibilities of showing different absorption in three different directions and hence show up to three colours. They are called pleochroic crystals. Isotropic crystals belonging to the cubic system do not show any colour change with their direction. Likewise, anisotropic crystals, when viewed along the optic axis, (normal to the circular section) also do not show any change, since there is no double refraction in the optic axis direction. The relative absorption of light also depends on the thickness of the section. Thicker crystals show greater absorption and deeper display of colours under the microscope. Some common minerals showing strong pleochroism are tourmaline, rutile, hornblende, piedmontite and cordierite.

5.7.9 Polarizing Microscope

The optical properties discussed above relate to transparent (non-opaque) crystals sliced into thin sections and can be studied with the help of a polarizing microscope, where the observations are made under polarized

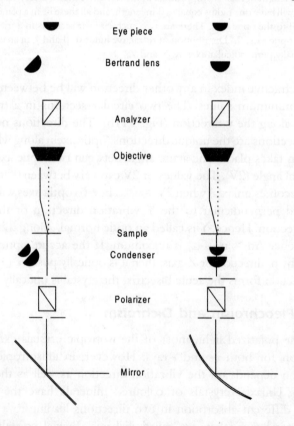

Fig. 5.12 A general layout of the optical path in polarizing: (a) orthoscopic and (b) conoscopic conditions.

light. The polarizing microscope is a compound microscope fitted with two polarizing devices. The one below the specimen is called the polarizer and the one above the specimen is called the analyzer. The general layout of the polarizing microscope is shown in Fig. 5.12. For more details readers should refer to books on Optical Microscopy.

Basically two types of observations are possible, namely: 1) one with parallel polarized light beam called orthoscopic conditions, and 2) converging polarized light beams called the conoscopic conditions. In the orthoscopic condition, the observations can be made either with the single polarizer (plane-polarized light) or with the two polarizers in crossed positions, i.e. their vibration directions are mutually perpendicular. The determination of the refractive indices of crystal and the study of pleochroism or dichroism are done under plane-polarized light.

The refractive indices of the crystals are measured under a polarized microscope using the liquid immersion technique. Here, a series of liquids of different refractive indices are matched with that of the crystal. When both the refractive indices of the crystal and the liquid are identical, one cannot visually distinguish the liquid from the crystal. The refractive index of such a matched liquid is independently measured with a refractometer which is equal to the refractive index of the crystal.

The phenomenon of pleochroism is observed when the stage of the microscope carrying the crystal is rotated. In this operation, the crystal is rotated with reference to the vibration of the polarizer. Thus, the change in colour in different crystal directions occurs because of the principle explained in section 5.7.7.

Observations under the crossed polarized light are used to study: 1) The extinction position, 2) Interference colours, and 3) Interference figures.

In the extinction position, a crystal becomes completely dark. This happens when the vibration directions of the crystal coincide with the vibration directions of the polarizer as well as that of the analyzer. In a complete rotation of 360 °, the extinction position can occur 4 times. In between the two extinction positions lies the maximum illumination. All isotropic crystals as well as anisotropic crystals viewed along the optic axis will remain dark under the crossed polarized light over the entire rotation of 360 °. This is because the vibration direction of the polarized light remains unchanged when transmitted through the crystal in such conditions. This is the simplest method of distinguishing between the isotropic and anisotropic crystals. If, in the extinction position of the polarizer, the vibration direction is parallel to the major crystal edges, it is called straight extinction. Crystals belonging to the tetragonal, hexagonal and orthorhombic systems will show perfect straight extinction in sections parallel to prism faces. Some sections cut parallel to pyramidal faces show symmetrical extinction. Other crystals of the monoclinic and triclinic systems invariably show inclined or oblique extinction.

Interference colours are very bright colours when observed in an anisotropic (birefringent) crystal under crossed polarized conditions. They are best seen in the maximum illumination position, i.e., the position 45° away from extinction position. The ordinary and the extraordinary rays emerging out of the crystal have a phase difference (because n_o is not equal to n_e). These phase-displaced waves enter the analyzer, and since both the waves are made to vibrate in the same plane, their interference is inevitable. When the phase difference is neither a whole wavelength $(0, 1\lambda, 2\lambda, \text{etc.})$ nor half wavelengths $(\lambda/2, 3\lambda/2, 5\lambda/2, \text{etc.})$, elimination of a certain spectral part of the white light takes place leading to the appearance of the complimentary colours. Depending upon the degree of phase difference, different orders of interference colours are possible. The interference colours also depend on the thickness of the crystal, the birefringence and the orientation. Introduction of accessories, such as a mica or a gypsum plate or a quartz wedge, can bring in a known degree of interference superimposing on the phase difference produced by the crystal additionally or subtractively. Such addition or subtraction of interference colours is used to determine the slow or fast vibration direction of the birefringent crystal. The order of interference colours are fixed with reference to the Newton's colour index chart.

5.7.9a   The Conoscopic Condition

In the conoscopic condition (Fig. 5.13a), the crystal section is observed under the convergent polarized light beam so as to observe the interference figure, and not the image of the crystal as observed under orthoscopic conditions. Because of the converging conditions of the light, the angle of convergence changes from the centre to the periphery of the crystal and will be more oblique from the centre to the periphery of the crystal. The o- and e-rays emerging from the crystal are also correspondingly oblique. In addition, the extent of splitting increases progressively. As a result, they interfere to form what are known as interference figures (Fig. 5.13b-e). In an uniaxial crystal, viewed along the optic axis, the interference figure appears with a dark cross superimposed on coloured concentric circular rings (Fig. 5.13c). The dark cross is called the isogyre, the intersection of isogyres is the emergence of optic axis. If it is a biaxial crystal viewed along an optic plane, two isogyres appears with concentric rings as shown in Fig. 5.13d & e, the centres of two isogyres represent the emergence of two optic axes. The interference figure obtained under conoscopic condition is used to determine the optic sign of

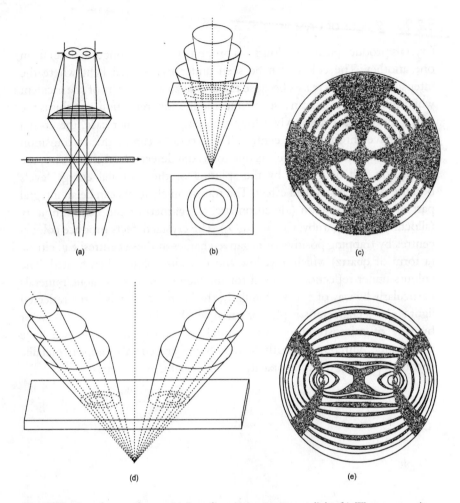

Fig. 5.13 Interference figures (a) Interference in convergent light (b) Waves cones in a uniaxial crystal (c) Uniaxial interference figure in convergent light (d) Wave cones in a biaxial crystal (e) Biaxial interference figure in convergent light

a crystal using accessory plates such as a mica plate, a gypsum plate or a quartz wedge. Since the vibration directions of slow and fast rays are known in these plates, they can be used to find whether the *e*-ray is slower (positive crystals) or faster than the *o*-ray (negative crystal). For details on the use of conoscopic figures, the reader may refer to a book on Optical Mineralogy.

5.7.9b   Colour of Crystals

Crystals produce pleasing colours—a property used to distinguish them from one another. The colour can be seen under transmitted light where the intensity of light decreases because the crystal absorbs a part of the incident white light. The crystal can also exhibit colour under reflection, and this is the only way of observation when they are opaque. In either case, the crystal should have chromophoric centres. These centres can be due to transition or rare earth elements, as well as due to crystal defects. In the first case, the absorption will be guided by the splitting of the d-orbital energy levels because of the crystal field effect. The transition elements can be an integral part of the composition (idiochromatic, e.g. garnet) or can be an impurity (allochromatic, e.g. ruby, Cr^{3+} in Al_2O_3). The crystal defects produce colour-centres by trapping positive or negative charges at these centres; e.g. citrine (a form of quartz) which is yellow and loses its colour when heated. The colours under reflection are used for the identification of opaque minerals particularly by way of streaks. Similarly, the lustre of crystals in the reflected light is also a characteristic phenomenon which is due to an additional factor by way of microstructural (or textural) modifications. Thus, we have adamantine, vitreous or pearly lustres. Some crystals also appear metallic (pyrite) indicating the total opacity to visible light, as in metals.

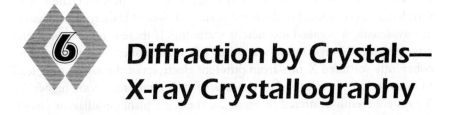

Diffraction by Crystals— X-ray Crystallography

6.1 INTRODUCTION

The internal structure of a crystal can be explored with any radiation whose wavelength should be comparable to the interplanar spacings, i.e. the spacings between the atomic planes with different Miller indices. These radiations can be from that part of the electromagnetic spectrum which have short wavelengths. Such short wavelengths are also possessed by fast moving subatomic particles, such as electrons or neutrons. The electromagnetic spectrum has been shown in Fig. 5.5 (Chapter 5). The interplanar spacings in most crystals are in the range of 1 to 10 Å, and in some rare cases extending up to 30 Å. The part of the electromagnetic spectrum having wavelengths corresponding to this distance are X-rays between the γ rays and deep ultraviolet rays. Similarly, the fast moving electrons or neutrons can also have this range of wavelengths, and they also can be used to probe into the internal structure of crystals. However, the interaction of an electron or neutron beam with the crystals is more involved because they are particles with the wave nature. Because of this and other practical reasons, these are less commonly used in studying crystal structures as compared to the use of X-rays. In this chapter, we proceed with the treatment of X-ray crystallography and deal qualitatively with the other two methods towards the end.

When such radiations of short wavelengths impinge upon a crystalline substance, two separate effects can be expected, namely absorption and scattering. Absorption of X-rays will result in excitation of the electrons from the inner orbitals of atoms contained in the crystal. This may even result in the ejection of these excited electrons out of the atoms. Sometimes, it can displace atoms from the regular lattices to produce vacancies. All these processes actually lead to the absorption of X-rays. A part of the absorbed X-rays can be re-emitted with decreased energies, i.e. a longer wavelength, giving rise to what is known as the fluorescent X-ray. Such absorption phenomenon is useful in the X-ray fluorescence analyses (XRF), for determining the chemical contents of a crystal.

The scattering of X-rays is the second phenomenon, wherein the X-rays interact with the electrons in an atom and, hence, with the least change in its wavelength as compared to the fluorescence of X-rays. If there is any change in wavelength, it is called incoherent scattering. If there is no change in the wavelength of X-rays during scattering, it is called coherent scattering. The coherently scattered X-rays from different electrons of the same atom, lead to the interference effects. The chances of such interferences are higher for X-rays coherently scattered by atoms of the same plane or adjacent planes.

This is analogous to the diffraction of light from an optical grating. The fine and closely spaced lines in a grating are two-dimensional equivalents to the atomic planes in the crystals. It is from this analogy that the coherent scattering of X-rays by crystal planes is called the X-ray diffraction.

It is interesting to note that though an optical grating can be imaged through an optical microscope, the same is not possible in the case of the so called "three-dimensional atomic grating", i.e. a crystal. The reason is that the change in the refractive index of X-rays is one in a million or less, which makes it impossible to construct a lens for the direct imaging of crystal structures with X-rays. We know that without a lens, the radiation cannot be bent to build an image. Therefore, in X-ray crystallography we only see the diffraction patterns. From these patterns we have to reconstruct the image of an internal atomic structure through mathematical calculations.

A crystal diffraction pattern should not be confused with the radiographic image of a human body, which is just a shadow generated from the difference in absorption of X-rays between the harder bones and softer tissues of the body.

6.2 Production and Properties of X-rays

When a high energy electron beam from the cathode of a vacuum tube [Fig. 6.1a] is made to strike at a positively charged anode (also called the target), the electrons can penetrate deep into the atoms of the target. They knock out electrons from the innermost K-shell, provided the energy of the impinging electron beam is sufficiently high. In such excited atoms, the electron vacancy of the inner K-shell is restored by the falling of electrons from the outer shells, i.e. *L*, *M*, etc. The probability of such an electron transfer is maximum from the immediate *L*-shell and to a lesser extent from the *M*-shell. Such transfers of electrons from the outer *L*, *M*, *N*-shells to the K-shell result in the emission of X-rays, whose energy is equal to the difference between the energies of the electronic shells (ΔE). The wavelength of the X-rays produced is

$$\lambda = ch / \Delta E \qquad (6.1)$$

where c is the velocity of light in a vacuum and h is the Planck's constant.

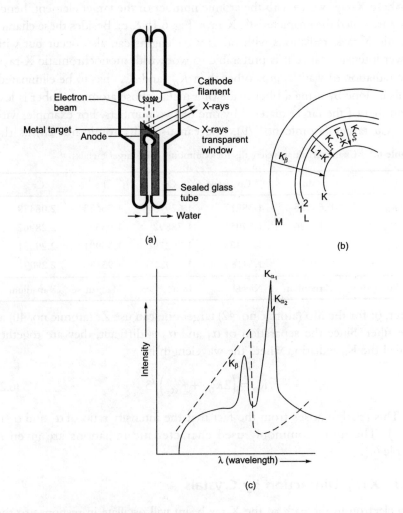

(a)

(b)

(c)

Fig. 6.1 (a) Section of the X-ray generation tube (b) Schematic energy level diagram illustrating the movement of electrons from M & L levels to K level (c) Schematic illustration of the intensities of K_{α_1}, K_{α_2} and K_{β} radiation. The dashed curve is the absorption by the filter so that K_{β} is eliminated.

Since the L-shells are made of two sublevels, L_1 and L_2, the X-rays produced from the transfer of electrons from these levels will have two close λ-values, α_1 and α_2. As they are produced from the transfer of electrons to the K-shell, they are called K_{α_1} and K_{α_2}. Likewise, the electron-transfer from the M to the K-shell will produce K_β radiation. The actual wavelength of these X-rays varies with the atomic number of the target element, hence they are called the characteristic X-rays (Fig. 6.1b & c). Besides these characteristic X-rays, radiations with other wavelengths can also occur but with lower intensity. Since it is preferable to work with monochromatic X-rays, the radiation of wavelengths other than K_{α_1} and K_{α_2} has to be eliminated. This is done by using a filter made of a metal whose atomic number is less than that of the target material by one or two numbers. For example, with the Cu target (atomic no. 29), one can use Ni (atomic no. 28) as the

Table 6.1   Characteristic wavelengths for commonly used target metals

λ	Mo	Cu	Co	Fe	Cr
K_β	0.63225	1.39217	1.62073	1.75653	2.08479
K_{α_1}	0.70926	1.54051	1.78892	1.93597	2.28962
K_{α_2}	0.71354	1.54433	1.79279	1.93991	2.29351
K_α	0.7107	1.5418	1.7902	1.9373	2.2909
Filter (foil)	Zirconium	Nickel	Iron	Manganese	Vanadium

filter, or for the Mo (atomic no. 42) target one can use Zr (atomic no. 40) as the filter. Since the separation of α_1 and α_2 is difficult, they are together called the K_α radiation where the wavelength is

$$K_\alpha = \left(2K_{\alpha_1} + K_{\alpha_2}\right)/3 \qquad (6.2)$$

This relation arises from the fact that the intensity ratio of α_1 and α_2 is 2 : 1. The more commonly used characteristic radiations are given in Table 6.1.

6.3   X-ray Diffraction by Crystals

An electron in the path of the X-ray beam will oscillate in response to the oscillatory changes in the electric field of the X-rays. This oscillation causes the electrons to emit X-rays in all directions with the same wavelength, thus the intensity of the impinging primary beam is slightly reduced. If we consider an atom in the path of the X-rays, then all the electrons in the atom can scatter the X-rays. The scattering power of the atom is proportional to the number of electrons in that atom.

The scattering of X-rays by crystals was first treated by von Laue. He considered a row of atoms periodically spaced at a translational distance 't' (Fig. 6.2a). Assume that a beam of X-rays is striking the row of atoms. The line p_1, p_2, p_3..., represents the plane, wherein each ray in the beam is in-phase. Then p_1, p_2, p_3 represents a plane wave front. The incident beam is at an angle, $\overline{\mu}_1$, to the row of atoms. As stated earlier, the X-rays are scattered in all directions by each atom. Of these scattered rays, let us consider a parallel set of scattered rays c_1, c_2 and c_3. These rays are observable only if they are travelling in-phase, which means that the path difference between each of these scattered rays c_1, c_2 and c_3 should be an integral multiple of the wavelength λ (i.e. $n\lambda$, where n is an integer). If the angle of scattering is $\overline{V}$, then the extra distance travelled by p_3 with respect to p_2 (Fig. 6.2b), is the path difference,

$$[r - s] = [t \cos \overline{V} - t \cos \overline{\mu}] = t\left(\cos \overline{V} - \cos \overline{\mu}\right) = n\lambda \qquad (6.3)$$

This is called the Laue equation. The diffraction occurs only if this condition is satisfied. To define space in three dimensions in a crystal, there should be three minimum intersecting lattice rows which are not in the same plane. Considering that the primary X-rays strike these rows at three different angles, μ_1, μ_2 and μ_3 and produce three different sets of diffracted beams at angles V_1, V_2 and V_3, there will be three consecutive Laue equations satisfying the diffraction conditions:

$$a\left(\cos V_1 - \cos \mu_1\right) = h\lambda$$
$$b\left(\cos V_2 - \cos \mu_2\right) = k\lambda \qquad (6.4)$$
$$c\left(\cos V_3 - \cos \mu_3\right) = l\lambda$$

Here a, b and c are the repeat distances of atoms in three rows and hence the unit cell parameters and h, k and l are the corresponding Miller indices.

Since the atoms scatter X-rays in all directions at the same time, the scattered rays maintain different angular relations for a given value of n in the Laue relation. We can therefore visualize a set of cones of diffracted rays centered around each row of atoms in a lattice. In fact, the configuration shown in Fig. 6.2b is a part of the cross-section of such a cone. As the value of the integer n (in Eq. 6.3) changes, the diffraction angle also changes, thus giving rise to concentric cones as shown in Fig. 6.2c. Since n can also have negative values, the cones are positioned on either side of the incident beam.

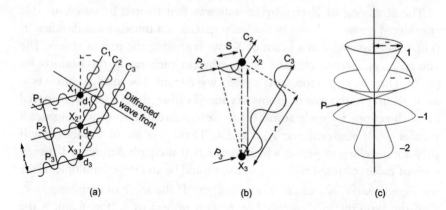

Fig. 6.2 (a) Laue's concept of the scattering of X-rays by a row of atoms (b) Blown-up sketch of the diffracted X-rays from points x_2 and x_3 to derive the Laue equation (see text) (c) Concentric cones of diffracted rays around a row of atoms where the value of angle $\bar{v}$ depends on the value of n in Laue's equation.

6.3.1 Bragg's Equation

W H Bragg and W L Bragg held that beams of diffracted X-rays can be treated in terms of reflections from the lattice planes of crystals. Consider two sets of planes A and B with an interplanar spacing of d (Fig. 6.3 a & b). If the X-ray beam of p_1, p_2 and p_3 is incident on these planes with the given angle θ_i, this gives rise to a reflected beam c_1 c_2 and c_3 with an angle θ_r (angle of reflection). When $\theta_i = \theta_r$, according to the law of reflection, the rays p_1, E_1 c_1 and p_2, E_2 c_2 should arrive at the same time at c_1, c_2, c_3, only if they are in-phase, which means, the path difference between p_1, E_1 c_1 and p_2, E_2 c_2 should be $n\lambda$ which is equal to $ME_2 + NE_2$

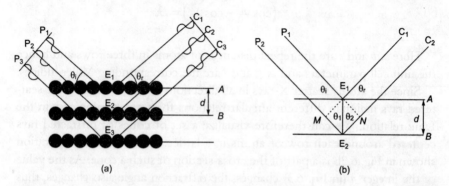

Fig. 6.3 (a) Bragg's conceptual diffraction of X-rays by parallel rows of atoms on a crystal lattice (b) Schematic consideration of X-ray reflection from two parallel planes to derive Bragg's equation for X-ray diffraction.

Considering the two right-angled triangles E_1ME_2 and E_1NE_2

$$\sin\theta_i = E_2M/E_2E_1 = E_2M/d \qquad (6.5)$$

Therefore,

$$E_2M = d\sin\theta_1$$

Similarly, in triangle E_1ME_2,

$$E_2N = d\sin\theta_2$$

Therefore, the total path difference $= 2d\sin\theta$, (since $\theta_1 = \theta_2$) which should be an integral multiple of λ,

i.e., $\qquad\qquad\qquad n\lambda = 2d\sin\theta$

The Braggs have illustrated this relation using NaCl crystal cut and polished parallel to the [111] plane. They demonstrated that the CuK_α beam reflects only when the incident angle, θ_i, has a definite angle of 13.7°, 28.27°, 45.27° and 71.3°. No reflections are noticed if θ_i has any other values. Thus, they explained the results in terms of the order of reflection wherein the value of 'n' changes from $n = 1$ (first order), 2 (second order), 3 (third order) or 4 (fourth order). These can also be considered as reflections from parallel layers of nh, nk and nl type reflections, i.e. (111), (222), (333) and (444) reflections. In fact, with CuK_α radiation, the reflection from the (555) plane could not be observed, because by substitution of the values of λ and d in [Eq. 6.6], we have

$$5 \times (1.548) = 2 \times (3.255)\sin\theta \qquad (6.7)$$

Therefore, $\sin\theta = 1.184$ (which is in excess of 1.00) and hence $\theta = 100.6°$, which is more than 90°. Thus, the fifth order reflection cannot be experimentally observed when $Cu\,K_\alpha$ is used. Whereas the same can be observed when MoK_α is used (See Table 6.1).

6.3.2 Ewald's Diagram

A graphical solution of Bragg's equation was first attempted by P P Ewald (Fig. 6.4a). He considered an imaginary sphere called the sphere of reflection, with the radius equal to $1/\lambda$. The crystal is placed at the centre of the sphere and a direct beam enters the sphere from the left at point A and emerges out of the sphere at O. Consider the NaCl crystal again, with the (111) plane parallel to the beam. At O, a perpendicular OP is drawn and equidistant points on this line correspond to the reciprocal distance of a set

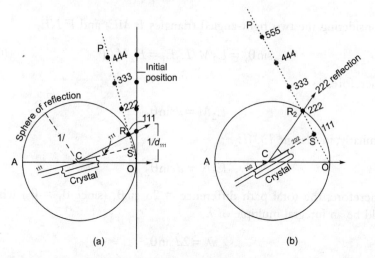

Fig. 6.4 (a) The NaCl crystal with the polished (111) face placed within an imaginary sphere called the Ewald sphere or sphere of reflection. The sphere radius is proportional to $1/\lambda$, where $\lambda = 1.5418$ Å for Cuk$_\alpha$ rays. Therefore, $1/\lambda = 0.65$ Å^{-1} units on the scale. At 0, where the beam ACO leaves the sphere, draw OP $\perp$ to dashed line CS. Thus OP is $\perp$ to the (111) plane. Along OP is located the reciprocal point (111), at R at a distance corresponding to $1/d_{(111)}$ from 0. This is 0.31 Å^{-1} units on scale because $d_{(111)} = 3.26$ Å for NaCl. Hence, point 111 is called the reciprocal lattice projection of the plane (111). The interplanar spacings of d$_{222}$ d$_{333}$ are one-half and one-third of d(111) respectively. Hence the points 222, 333 reflection plot 2 and 3 times as far as 111 from 0. (b) The crystal rotated about an axis normal to the page at C such that OP is also rotated to maintain its perpendicularity to (111). When 222 reflection touches the sphere, a diffracted beam called 222 'reflection' travels from C toward point 222. Thus, whenever a *hkl* reflection touches the Ewald sphere the Bragg's condition is satisfied.

of parallel crystallographic planes (in this case $1/d(111)$). It can be seen that $2/d(111) = 1/d(222)$ or $3/d(111) = 1/d(333)$, and so on.

Consider that the crystal is rotated around an axis perpendicular to the plane of the paper, then the line OP also correspondingly rotates. Fig 6.4b is the rotated second position.

In Fig. 6.4b, as the point R touches the outer periphery of the sphere, a diffracted beam CR leaves the sphere. At this position, the angle of incidence for NaCl is 13.7°. The general solution for Bragg's equation can be demonstrated as follows.

In the triangle COR, CR = CO = $1/\lambda$. The line CS divides the triangle COR into two right-angled triangles. Also CS is the extension of the (111) plane from the crystal. Under these conditions, SR = SO. From the right-angled triangle, CSR,

$$\sin\theta_{(111)} = SR / CR = 1 / 2d_{(111)} / (1/\lambda) = \lambda / 2d_{(111)} \qquad (6.8)$$

$$\therefore \lambda = 2d_{(111)} \sin\theta_{(111)}$$

This is the restatement of Bragg's equation (6.6) for $n = 1$. If the crystal is rotated further such that angle $\theta_2 = \theta_{(222)}$, the line OP rotates further. When the point 222 touches the sphere, the reflected beam leaves the sphere of reflection at R_2. Here again, it can be shown that $\sin\theta_{(222)} = \lambda/2d_{(222)}$. In general, a reflected beam leaves the sphere only when the other points (333) or (444) touch the sphere as the crystal is rotated. It can be seen that the point corresponding to (555) does not touch the sphere of reflection of $1/\lambda = 1/1.5418$, which is taken as the radius of the sphere. If on the other hand, MoK_α (= 0.7107 Å) is used, it is possible to observe the (555) reflection because the radius of the Ewald's reflection sphere [$=1/\lambda$] is larger. This construction is valid notwithstanding whether the interaction of X-rays with the crystal is treated in terms of scattering as conceived by Laue, or reflection as envisaged by Bragg.

6.3.3 The Concept of Reciprocal Lattices

All the X-ray diffraction data, irrespective of the experimental methodology, are recorded on the reciprocal distances. This is a consequence of Ewald's sphere of reflection, where the reciprocal distances ($1/\lambda$ or $1/d$) are considered. In order to understand this concept, let us consider a two-dimensional lattice (i.e., a lattice net). The origin of this net is located at the point O in Ewald's sphere. We plot the points outward from the origin corresponding to $1/d_{[hkl]}$ for each set of planes (Fig. 6.5). If all the (100, 010, 001, etc.) planes and the corresponding higher order reflections (200, 300 or 020, 030, 002, 003, etc.) are reciprocally plotted, the result is a set of lattice points known as the reciprocal lattice. This is illustrated in Fig. 6.5a taking the *ab* plane of a monoclinic lattice, wherein *c* is perpendicular to the plane of the paper. In the direct lattice, the 010 plane is *b* distance away from the origin. Likewise, (020), (030), and (040) are parallel planes at $1/2b$, $1/3b$, $1/4b$ times away from the origin. Draw a perpendicular to this set of planes from the origin outward (OB) as shown in Fig. 6.5a. The first point on that line will be $1/d_{(010)}$, the next will be $1/d_{(020)}$, etc. so that 010, 020 and 030, etc. successively appear as one moves away from the origin.

Now consider the h00-type planes (100, 200, 300, etc.) and draw perpendiculars to these. The reciprocal points appear again on this line corresponding to 100, 200, 300 as one moves away from the origin. Let us take a third set of planes (110, 120, 130, etc.) which are non-basal, as shown in Fig. 6.5b. Unlike in the earlier two cases, each of these planes have perpendiculars drawn from the origin. The reciprocal distance corresponding to $d_{(120)}$, $d_{(130)}$, etc. will lie on the straight line which is the 0k0 reciprocal line. Similarly, the other types of reflections, 210, 220, 230, etc. will lie on the

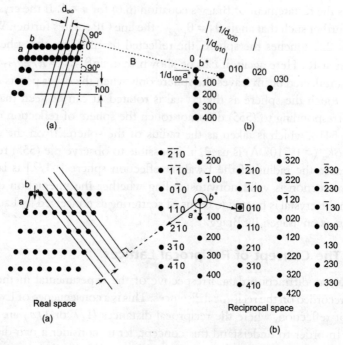

Fig. 6.5 The relation between (a) the direct (real) lattice in real space and (b) the reciprocal lattice in the reciprocal space. The spacings of (100) (010) planes perpendiculars drawn to the (010) (100) planes intersect at • which is the origin of reciprocal lattice. Each point along a row represents one set of parallel plane in real lattice and the repeat distance is reciprocal of real distance in real space, e.g., b* = 1/d$_{010}$ a* = 1/d$_{100}$.

same plane as shown. The interconnection of these reciprocal points leads to a two-dimensional clinonet indicating that the original symmetry and, hence, the angular relations between the axes are preserved.

In general, when any crystal is incidental with the X-ray beam, one can imagine that: 1) the crystal is centred within the sphere of reflection of radius 1/λ, 2) the reciprocal lattice has its origin at the point '•' where the direct beam leaves the sphere, and 3) as the crystal is rotated, the reciprocal lattice correspondingly rotates. As each point successively touches the circumference of the sphere, a diffraction beam emerges out of the sphere. The diffraction patterns recorded, therefore, are in the reciprocal lattice. The unit cell parameters in an orthogonal crystal *a, b* and *c*, have the reciprocals *a** [=1/d(100)], *b**[=1/d(010)] and *c**[=1/d(001)]. In the case of systems derived from the clinonet, namely monoclinic and triclinic systems, as well as in the trigonal and hexagonal systems, the angularity of the axes has to be taken into consideration. The relationship between direct and

Table 6.2 Dimensional relationships between the direct and reciprocal triclinic unit cell

---

Interaxial Angles

$$\cos\alpha = \frac{\cos\beta^*\cos\gamma^* - \cos\alpha^*}{\sin\beta^*\sin\gamma^*}$$

$$\cos\alpha^* = \frac{\cos\beta\cos\gamma - \cos\alpha}{\sin\beta\sin\gamma}$$

$$\cos\beta = \frac{\cos\alpha^*\cos\gamma^* - \cos\beta^*}{\sin\alpha^*\sin\gamma^*}$$

$$\cos\beta^* = \frac{\cos\alpha\cos\gamma - \cos\beta}{\sin\alpha\sin\gamma}$$

$$\cos\gamma = \frac{\cos\alpha^*\cos\beta^* - \cos\gamma^*}{\sin\alpha^*\sin\beta^*}$$

$$\cos\gamma^* = \frac{\cos\alpha\cos\beta - \cos\gamma}{\sin\alpha\sin\beta}$$

$$\frac{\sin\alpha}{\sin\alpha^*} = \frac{\sin\beta}{\sin\beta^*} = \frac{\sin\gamma}{\sin\gamma^*}$$

Cell Edges

$$a_o = \frac{b^*c^*\sin\alpha^*}{V^*} = \frac{1}{d_{(100)}^*}$$

$$a^* = \frac{b_o c_o \sin\alpha}{V} = \frac{1}{d_{(100)}}$$

$$b_o = \frac{a^*c^*\sin\beta^*}{V^*} = \frac{1}{d_{(010)}^*}$$

$$b^* = \frac{a_o c_o \sin\beta}{V} = \frac{1}{d_{(010)}}$$

$$c_o = \frac{a^*b^*\sin\gamma^*}{V^*} = \frac{1}{d_{(001)}^*}$$

$$c^* = \frac{a_o b_o \sin\gamma}{V} = \frac{1}{d_{(001)}}$$

Cell Volumes

$$V = a_o b_o c_o \sqrt{1 - \cos^2\alpha - \cos^2\beta - \cos^2\gamma + 2\cos\alpha\cos\beta\cos\gamma}$$

$$= a_o b_o c_o \sin\alpha\sin\beta\sin\gamma^*$$

$$= a_o b_o c_o \sin\alpha\sin\beta^*\sin\gamma$$

$$= a_o b_o c_o \sin\alpha^*\sin\beta\sin\gamma$$

$$V^* = a^*b^*c^* \sqrt{1 - \cos^2\alpha^* - \cos^2\beta^* - \cos^2\gamma^* + 2\cos\alpha^*\cos\beta^*\cos\gamma^*}$$

$$= a^*b^*c^* \sin\alpha^*\sin\beta^*\sin\gamma$$

$$= a^*b^*c^* \sin\alpha^*\sin\beta\sin\gamma^*$$

$$= a^*b^*c^* \sin\alpha\sin\beta^*\sin\gamma^*$$

Table 6.3　Interplanar Spacing d_{hkl} as related to reflection indices and unit cell dimensions

Crystal System	$\dfrac{1}{d_{hkl}^2}$
Isometric	$\dfrac{1}{a_0^2}\left(h^2 + k^2 + l^2\right)$
Tetragonal	$\dfrac{h^2 + k^2}{a_0^2} + \dfrac{l^2}{c_0^2}$
Orthorhombic	$\dfrac{h^2}{a_0^2} + \dfrac{k^2}{b_0^2} + \dfrac{l^2}{c_0^2}$

Hexagonal	
Hexagonal indices	$\dfrac{4}{3a_0^2}\left(h^2 + hk + k^2\right) + \dfrac{l^2}{c_0^2}$
Rhombohedral indices	$\dfrac{1}{a_r^2}\dfrac{\left(h^2 + k^2 + l^2\right)\sin^2\alpha + 2\left(hk + kl + lh\right)\left(\cos^2\alpha - \cos\alpha\right)}{1 - 2\cos^2\alpha + 3\cos^2\alpha}$

Monoclinic	
	$\dfrac{\dfrac{h^2}{a_0^2} + \dfrac{k^2}{b_0^2} - \dfrac{2hk\cos\gamma}{a_0 b_0}}{\sin^2\gamma} + \dfrac{l^2}{c_0^2}$　(first setting)
	$\dfrac{\dfrac{h^2}{a_0^2} + \dfrac{l^2}{c_0^2} - \dfrac{2hl\cos\beta}{a_0 c_0}}{\sin^2\beta} + \dfrac{k^2}{b_0^2}$　(second setting)

Triclinic	
	$\left[\dfrac{h^2}{a_0^2}\sin^2\alpha + \dfrac{k^2}{b_0^2}\sin^2\beta + \dfrac{l^2}{c_0^2}\sin^2\gamma + \dfrac{2hk}{a_0 b_0}(\cos\alpha\cos\beta - \cos\gamma) \right.$ $\left. + \dfrac{2kl}{b_0 c_0}(\cos\beta\cos\gamma - \cos\alpha) + \dfrac{2lh}{c_0 a_0}(\cos\gamma\cos\alpha - \cos\beta)\right]$ $+ \left[1 - \cos^2\alpha - \cos^2\beta - \cos^2\gamma + 2\cos\alpha\cos\beta\cos\gamma\right]$

reciprocal lattice constants is given in Table 6.2. The angles between axes in the reciprocal lattice are denoted by α^*, β^* and γ^*.

The interrelation between the interplanar spaces (*d*-spacings) to the Miller indices (*hkl*) and the unit cell parameters follow the relation shown in Table 6.3.

6.4 METHODS IN X-RAY CRYSTALLOGRAPHY

In terms of the specimen handled, two methods can be identified, namely, the powder method and the single crystal method. In the former, the specimen is a collection of crystallites. Since these fragments are completely randomly oriented, the incident X-ray beam meets with every possible lattice plane, oriented in all directions. Whereas in the single crystal methods, the whole specimen is a single piece without any discontinuity in the lattice arrangements. Single crystal does not mean a big crystal; as with modern instruments, even a small crystal of about 20μm size can be used for single crystal analyses.

6.5 THE POWDER METHOD

The powder method essentially has two different ways of registering the diffracted X-rays: 1) The whole diffraction pattern is recorded simultaneously on a photographic film called the powder photographic method, 2) The diffraction pattern is scanned by a counter device, (Geiger proportional or scintillation counter) or a solid state semiconductor detector. The counter or detector registers the diffracted beam in successive stages, away from the direct beam.

6.5.1 Photographic Method

Amongst the different photographic methods, the Debye–Scherrer is the most commonly used. The configuration of the powder specimen and the film in a Debye–Scherrer method is shown in Fig. 6.6.

A few milligrams of fine powder sample is mixed with a glue, such as collodian, and shaped into a cylindrical body (about 0.5 mm dia and 5–10 mm long). The powder can also be filled in a glass capillary made of lead-free borate glass. The specimen is mounted at the center with its axis coinciding with the centre of the cylindrical film. The Debye–Scherrer camera shown in Fig. 6.6 has a collimator as well as the beam-trapper. The film is mounted tight against the inner cylindrical wall of the camera. It is appropriately punched with holes for the collimator and the beam trapper to

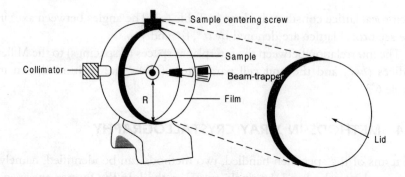

Fig. 6.6　The Debye–Scherrer powder diffraction camera will be closed with a light-proof lid after fixing the film.

penetrate. This method of film mounting is called the Straumanis mounting. The light-proof lid prevents leakage of extraneous light. It is obvious that the mounting and removal of the film is done in the dark.

As already mentioned, in a powder sample, several random crystallites are exposed to the X-ray beam, with all the lattice planes oriented so as to satisfy Bragg's conditions for diffraction, $n\lambda = 2d\sin\theta$. Irrespective of the orientation of crystallite, the given hkl plane will diffract X-rays to produce one cone. It is the net contribution from all the crystallites for particular hkl value and the direction of the beam. Diffraction of X-rays from different planes of varying d-spacings would give rise to several concentric cones of varying 2θ as shown in Fig. 6.7a. This is possible not only in the forward reflections (i.e. 2θ varying from 0 to 90°), but also for the back reflections ($2\theta = 90$ to 180°). Because the photographic film is in the form of a strip,

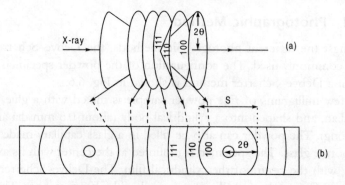

Fig. 6.7 (a) The illustration of diffraction cones emerging from a powder sample, each cone represents reflection from a particular (hkl) plane (b) A film exposed in the Debye–Scherrer camera, a pair of the concentric arcs represent each cone.

each cone produces a pair of arcs and the distance between the corresponding arcs is proportional to 4θ.

A powder sample in a Debye–Scherrer camera method is normally exposed for a couple of hours in order to obtain a good diffraction pattern. The film, when developed after exposure, would appear as shown in Fig. 6.7b. The distance between the two corresponding arcs from the film strip will be proportional to the diffraction angle. In Fig. 6.7b, if S is the distance between the two arcs of a cone, then the angle of diffraction is $θ = S/4R$, where R is the radius of the camera. If R is chosen as a multiple or submultiple of the radian (i.e. $180/π = 57.3$, a factor to convert reading from linear distances to degrees), then S will directly give the angle of diffraction in degrees, i.e. $θ = (S/4R)(180/π)$. If R is 57.3, then 1 mm is 1/2 degree, if R is 57.3/2, then 1 mm is 1 degree.

The distance between the pair of arcs obtained from the back reflection corresponds to the diffraction angle, $2θ = 180 - (S/2)$. Another piece of useful information that we derive from the powder photograph is the difference in the relative intensities of different reflections on the photographs, which can be quantified using an optical densitometer.

6.5.2 Powder Diffractometer

The tedious steps involved in the photographic method, namely the long duration of exposure, developing the film in dark room and problems such as the shrinkage of film and the measurement of relative intensities, have been overcome by the diffractometer technique. The diffractometer recording has the flexibility of achieving a higher resolution by way of recording small ranges of $2θ$ on an expanded scale. The diffractometer essentially contains a Geiger–Muller counter/scintillation counter or a solid state semi-

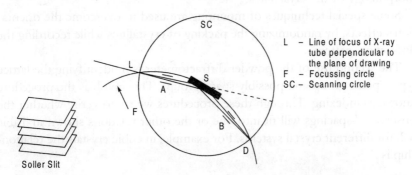

Fig. 6.8 Illustration of parafocussing in a powder diffractometer, A & B Soller slits; S-sample, D-detector.

conductor detector for measuring the intensity of the diffracted beam. The geometry of mounting the slit, the sample and the counter is shown in Fig. 6.8 which is based on a parafocussing arrangement. The powder sample is smeared as a thick film on a glass slide using any one of the organic polymers as binder, which, in turn, is mounted at the centre of the goniometer. The motor-driven gear rotates the goniometer and the glass slide by an angle θ and at the same time the X-ray detector is moved by an angle 2θ. Hence, the movements are coupled. The signals received by the counter, suitably amplified, are recorded on a strip-chart recorder. Each (*hkl*) reflection on the recording appears as a peak whose height above the background indicated the intensity of the diffracted beam. Since the scanning speed of the detector and the speed of the chart recorder can be independently varied, it is possible to achieve the desired resolutions of the diffraction pattern. Whereas the distance between the two arcs in a powder photograph corresponds to 4θ, the angle in a diffractometer recording corresponds to the 2θ values. Figure 6.9 shows the comparison of a photograph and a diffractogram.

6.5.3 Interpretation of Powder Patterns

The diffracted angles measured are converted into *d*-spacings using Bragg's relation, $d = \lambda/2d \sin\theta$, where λ is the wavelength of the radiation.

The *d*-spacings are listed in a table in a decreasing order of values. Three or four of the most intense reflections are normally used for identifying the unknown crystals. This is because no two similar crystalline substances have identical *d*-spacings with the same relative intensities. However, the relative intensities can sometimes be affected due to the preferred orientation of crystallites during mounting, because of flaky cleavages and the sheet-like morphology of the crystallites.

Some special techniques of mounting are used to overcome the orientational effects, by randomizing the packing of crystallites while recording the diffraction pattern.

The second use of the powder diffractometer is in identifying the lattice types (point group and possibly space group). This involves the procedure known as indexing. Through these procedures we try to verify whether the observed *d*-spacings will fit into one or the other relations shown in Table 6.2, for different crystal systems. For example, in cubic crystals the relationship is

$$d(hkl) = a \Big/ \sqrt{\left(h^2 + k^2 + l^2\right)} = a/\sqrt{N} \quad \text{where } N = h^2 + k^2 + l^2 \quad (6.9)$$

There are two ways to check whether this relation is valid for every observed *d*-spacing: 1) the analytical method, and 2) the graphical method.

The analytical procedure is illustrated with the powder pattern of spinel
listed in Table 6.4. Here the *d* values appear in the decreasing order and are
designated as d_1, d_2, d_3, etc. All the *d* values are squared. The largest value, i.e.,
$(d_1)^2$ is divided by $(d_2)^2$, $(d_3)^2$, $(d_4)^2$, etc. The numbers so obtained have to be
rounded off to the next integer. If that is not possible, then these ratios are
multiplied by a common integer, so that all the ratios will be close to a
whole number. In Table 6.4 it is chosen as 3. Thus, in column 4 we get
whole numbers like 3, 8, 11, 16, etc. For the cubic system, this corresponds

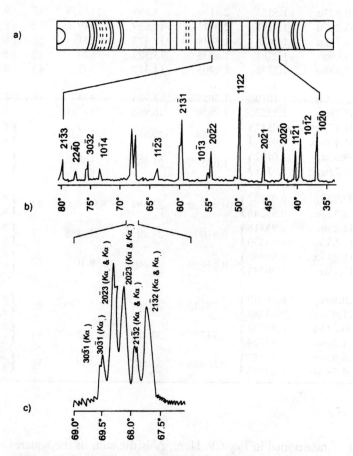

Fig. 6.9 Comparison of a powder photograph with a diffractogram (a) Powder photograph
of quartz with Cu k_α radiation (b) Diffractogram between angles 35° and 80° (c) Higher
resolution of closely appearing diffraction peaks by increasing the chart speed. The k_{α_1} & k_{α_2}
peaks are fully resolved. $hk\bar{i}l$ (where $h + k = -i$) indices are shown because quartz belongs to
hexagonal system.

Table 6.4   Observed 2θ and calculated d_1^2/d^2 values for a powder film of synthetic spinel, $MgAl_2O_4$

$2\theta_{obs}$	d	1 d^2	2 d_1^2/d^2	3 hkl	4 N^*	5 a_{calc}^{**}
19.02α	4.6658	21.7697	1	111	3	8.0814
31.30α	2.8577	8.1664	2.666	220	8	8.0828
36.894α	2.4365	5.9365	3.667	311	11	8.0810
44.84α	2.0212	4.0852	5.329	400	16	8.0848
55.702α	1.6501	2.7232	7.994	422	24	8.0838
59.427α	1.5553	2.4190	8.999	511/333	27	8.0816
65.287α	1.4291	2.0423	10.659	440	32	8.0812
68.736α	1.3656	1.8649	11.673	531	35	8.0790
74.19α	1.2781	1.6335	13.327	620	40	8.0834
77.388α	1.2331	1.5205	14.317	533	43	8.0860
82.691α	1.1670	1.3619	15.984	444	48	8.0852
85.78α	1.1327	1.2830	16.968	711/551	51	8.0891
94.171α₁	1.0517 ⎱	1.1063	19.678	731/553	59	⎰ 8.0783
94.451α₂	1.0520 ⎰					⎱ 8.0806
99.37α₁	1.0102 ⎱	1.0203	21.337	800	64	⎰ 8.0816
99.72α₂	1.0101 ⎰					⎱ 8.0808
107.92α₁	0.95257 ⎱	0.90729	23.994	822/660	72	⎰ 8.0828
108.327α₂	0.95248 ⎰					⎱ 8.0821
111.26α₁	0.93318 ⎱	0.87094	24.995	751/555	75	⎰ 8.0816
111.656α₂	0.93330 ⎰					⎱ 8.0826
116.944α₁	0.90365 ⎱	0.81640	26.665	8.40	80	⎰ 8.0825
117.45α₂	0.90345 ⎰					⎱ 8.0807
120.50α₁	0.88719 ⎱	0.78582	27.667	911/753	83	⎰ 8.0827
121.074α₂	0.88686 ⎰					⎱ 8.0797
130.744α₁	0.84733 ⎱	0.71797	30.321	931	91	⎰ 8.0836
131.366α₂	0.84734 ⎰					⎱ 8.0831
138.028α₁	0.82497 ⎱	0.68056	31.987	844	96	⎰ 8.0830
138.785α₂	0.82495 ⎰					⎱ 8.0828

* $N = h^2 + k^2 + l^2$
** $a = d\sqrt{N}$

to the N mentioned in Eq. 6.9. Here, N is the sum of the squares of three whole numbers. For example, $N = 3$ means $[(1)^2 + (1)^2 + (1)^2]$ or $N = 8$ means $[(2)^2 + (2)^2 + (0)^2]$, and $N = 11$ means $[(3)^2 + (1)^2 + (1)^2]$, and so on. We are able to assign the (hkl) indices for each of the reflections. If the assignment is correct, the cell parameter $a = d\sqrt{N}$ for the cubic system

Table 6.5 X-ray powder diffraction data of sodium fluoride

$D*$ cm	$(\pi-2\theta)**$ rad	θ deg	$\sin^2\theta$	hkl	$h^2 + k^2 + l^2$
22.1546	141.04	19.48	0.1112	200	4
19.4371	123.74	28.13	0.2223	220	8
17.6307	112.24	33.88	0.3335	222	12
15.1425	96.40	41.80	0.4446	400	16
13.1319	83.60	48.20	0.5558	420	20
11.0741	70.50	54.75	0.6669	422	24
6.1041	38.86	70.57	0.8892	440	32

should be single-valued for all d-values observed leading to the same a-values as shown in the last column. If the crystal was not cubic then we will not get a single value of a, because the ratios of $d_{(1)}^2/d_{(2)}^2$, etc. cannot be rounded off to the nearest integer even after multiplication by a common whole number. Furthermore, the forbidden values such as 7, 15, 23, etc. will also appear, which cannot be expressed as the sum of three squared integers $[(h)^2 + (k)^2 + (l)^2]$.

Graphical Solution

In the graphical method of indexing, we again illustrate with a cubic crystal of NaF. Eq. 1 can be written as

$$\sin^2\theta = \frac{\lambda^2}{4a^2}\left[\left(h^2\right)+\left(k^2\right)+\left(l^2\right)\right] \tag{6.10}$$

$$\sin^2\theta = \frac{\lambda^2}{4a^2}N = KN \tag{6.11}$$

Where K is a constant $= \lambda^2/4a^2$. Therefore, $\sin^2\theta$ versus N is a straight plot, passing through the origin.

A graph with $\sin^2\theta$ on the y-axis and N on the x-axis is constructed. Initially, the value of N corresponding to a given $\sin^2\theta$ is unknown. In order to circumvent the difficulty, the observed $\sin^2\theta$ are marked by way of a horizontal line as shown in Fig. 6.10. The N values are represented by vertical lines. If a ruler is pivoted on the origin and rotated until all the

---

* D is the distance between the members of each pair of lines in the Debye–Scherrer photograph from a camera of diameter, R = 4.5 cms

** $\dfrac{D}{2R} \times 57.3$

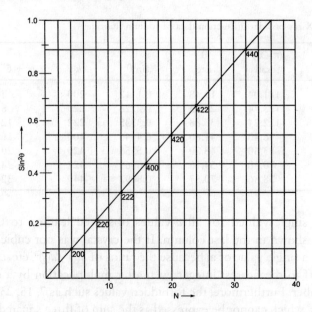

Fig. 6.10   sin²θ vs N-plot for indexing the X-ray reflections of sodium fluoride (cubic system)

intersection points of the horizontal and vertical axes are cut by the ruler, the N value of the corresponding sin²θ are read on these intersection points. Table 6.5 lists the sin²θ for NaF and the N values are read from Fig. 6.10. From the N values, one can obtain the *hkl* as well as the unit cell parameter. The latter can also be calculated from the slope of the graph, which is equal to $\lambda^2/4a^2$.

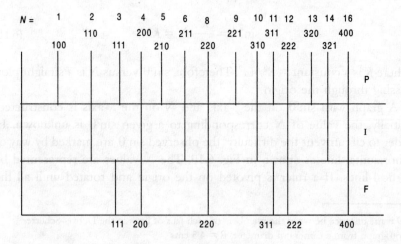

Fig. 6.11   Miller indices of reflections from a cubic crystal. P = primitive, I = body centred, F = face centred.

In the cubic system, the Miller indices can be used for distinguishing the Bravais lattice. Here, N for the reflection arriving from the P-type lattice (primitive) will appear in the order 1, 2, 3, 4, 5, 6, 8, etc. It is evident from Fig. 6.11 that the first reflection corresponds to (100) (i.e. $N = 1$) followed by (110), (111), (200), (211), etc. For the body-centred cubic crystals (I type), the first reflection corresponds to (110) (i.e. $N = 2$), followed by (200), (211), etc. In general, $[h^2 + k^2 + l^2]$ is even for the body-centred lattice. For the face-centred lattice (F type), the first reflection is (111) (i.e. $N = 3$) followed by (200), (220), etc. In general, for the F-type lattice, (h, k, l) values are either all even or all odd. That is, there are no mixed (odd plus even) (hkl) values. This method of identifying the Bravais lattice type is extendable to other crystal systems. Such restrictions in the indices are called systematic absences and are widely used in X-ray crystallography. It may be mentioned here that the splitting of N and assigning of the hkl value is equivocal, i.e. if $N = 1$, then it can be (100) or (010) or (001); or if $N = 8$, then it can be 220, 022 or 202. Similarly, some numbers like 9 can be 300 or 221 or $N = 27$ can be 511 or 333. Hence, indexing by using the powder pattern remains tentative and needs to be substantiated with single crystal studies.

Indexing procedures of crystals of lower symmetry exist, though they are more involved because of the higher number of variables, as may be seen from the relations given in Table 6.2. Excellent treatises on this subject exist by way of books on X-ray crystallography, a few of which are listed at the end of this book.

Yet another use of the X-ray powder pattern is in determining the accurate unit cell parameters. Some of the contributory error functions are absorption error, eccentricity error and inaccuracy of the goniometer or powder-camera diameter. The absorption error can be overcome by using the Nelson–Reiley function, $1/2[(\cos^2\theta/\sin\theta + \cos^2\theta/\theta)]$ plotted against the observed cell parameters. The extrapolation of this function to $\theta = 90°$ will produce a high accuracy for the cell parameter.

Accurate cell parameters are very useful in determining the chemical composition of the phases of solid solution series, such as $Mg_2SiO_4 - Fe_2SiO_4$ or $Fe_3Al_2(SiO_4)_3 - Mg_3Al_2(SiO_4)_3$, etc. If the sample contains two or more phases, the relative intensities of their characteristic reflections can be made use of in the quantitative estimation of the individual phases. The X-ray powder patterns are also useful in studying textures such as grain orientation or even predominantly platy character, etc., particularly in metals and alloys.

6.6 SINGLE CRYSTAL TECHNIQUES

The use of single crystals is essential in X-ray crystallography because it provides the means for direct measurement of cell parameters. Once the cell parameters are known, the indexing of the diffraction patterns becomes extremely simple; besides, determination of the system becomes easier. The unequivocal assignment of *hkl* values is possible unlike in the powder method. Not only can the reflections such as (020) and (200) be separated, their individual intensities can also be determined which was impossible in the powder method. Depending upon the movement involved for the crystal or the recording device (film or counter), two categories of single crystal methods exist. One is the Laue method, where both the crystal and the recording film remain stationary. In the other methods, the crystal invariably moves. In the rotation as well as oscillation method, the film is stationary. In the Weissenberg and Precession techniques, both the crystal and the film are moving.

6.6.1 Laue Method

The stationary single crystal is incident with white X-rays. Therefore, it has a range of wavelengths (See Section 6.2), as the ray are unfiltered. The Laue camera contains two flat photographic plates with a goniometer in the middle, on which the stationary crystal is mounted. The back-reflection plate will have a hole through which a collimater is inserted and the front-reflection plate carries a beam trapper. Since the crystal is stationary, the reciprocal lattice points are also stationary. Because the radiation used is

 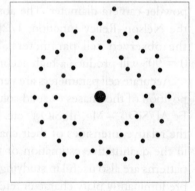

Fig. 6.12   (a) The Laue camera with cassettes for recording front and back reflections. (b) Sketch of a typical Laue pattern showing 4-fold symmetry.

white X-rays, there will be numerous wavelengths and each of them produce a sphere of reflection of its own. Therefore, each of the reciprocal
lattice points will be intersected at least by one of these spheres of reflections, so that the diffraction spots are generated. The Laue photograph
consists of a series of spots. If the X-ray beam is incident, parallel to the
symmetry axis of a crystal, the resulting Laue photograph will display the
symmetry of that axis. The Laue photographs are useful for directly finding
out the axis of symmetry in a crystal.

Figure 6.12a shows the Laue photographic arrangement and 6.12b a
typical example of reflection along a 4-fold axis. The absence of spots towards the central region in a Laue photograph is because of the sharp cut-
off of white X-ray radiation at shorter wavelengths, and hence the absence
of reflections by the small Bragg's angle of θ. Although the axis of the
system can be distinguished, the presence of the centre of symmetry or the
mirror plane cannot be deciphered from the Laue photographs. Consequently, in a Laue picture, all the non-centrosymmetric point groups also
show an apparent centre of symmetry. As a result, the 32 point groups
reduce to 11 Laue groups (see Table 6.6). The lightly-written notations are
the ones which are hidden. These are also called the Friedel classes and have
relevance to some of the physical properties of crystals. For example, the

Table 6.6 The eleven Laue groups (or Friedel classes)

TRICLINIC

$\bar{1}$,	1

MONOCLINIC			ORTHORHOMBIC		
2/m,	2,	m	2/m 2/m 2/m,	222,	mm2

TETRAGONAL

4/m,	4,	$\bar{4}$	4/m 2/m 2/m,	422,	4mm,	$\bar{4}$ 2m

HEXAGONAL

6/m,	6,	$\bar{6}$	6/m 2/m 2/m,	622,	6mm,	6m2

HEXAGONAL (TRIGONAL SUB-SYSTEM)

$\bar{3}$,	3	$\bar{3}$ 2/m,	32,	3m

ISOMETRIC OR CUBIC

2/m $\bar{3}$,	2 3	4/m $\bar{3}$ 2/m,	432,	$\bar{4}$ 3m

The non-centrosymmetric classes are shown in light print. If the centre of symmetry is added
to them they become the centrosymmetric class (shown in bold print). When the centrosymmetric
property is measured on non-centrosymmetric crystal, it responds with a pseudo-effect as if it
possessed the center of symmetry.

thermal expansion of a non-centrosymmetric crystal exhibits an apparent centre   of symmetry. Indexing of individual spots in a Laue photograph is difficult, because of the uncertainity in the wavelength of X-rays from which a given spot has originated. Although the use of monochromatic rays can overcome this problem, this reduces the number of diffracted spots tremendously. The most important use of Laue photographs in crystallography is in orienting the crystals to a great accuracy. The symmetrical disposition of the spots around the central spot is extremely sensitive to the slightest inclination of the axis along which the crystal is oriented.

6.6.2 Rotation Photograph Method

Here, the crystal mounted on a goniometer is rotated around an axis by 360°. The cylindrical film surrounds the crystal. The monochromatic X-rays are used here. The cones of reflection for different orders of reflections (n) will be coaxial with the axis of rotation, which is also the axis of the cylindrical film. These reflection cones intersect the film to produce spots which lie on parallel lines when the film is unfolded (Fig. 6.13). They are called the layer lines. Assume that the crystal is mounted on the c-axis. The layer line which coincides with the central hole is the zero-layer line or equatorial line. The subsequent upper and lower layers are designated as first, second or third layer lines. The layer lines are related to the repeat distances which is the present case will be the repeat distance along the c-axis (c-parameter). If H is the layer separation distance and R is the radius of the camera, then $H/$

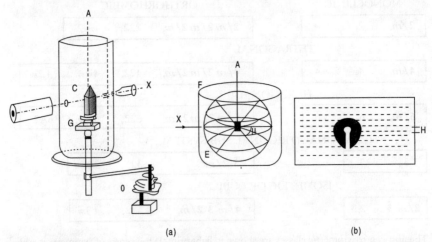

(a)   (b)

Fig. 6.13   (a) Rotating camera arrangement. E = Ewald sphere, F = film, A = axis of rotation, X = X-ray, O = oscillation cams, G = goniometer. (b) Typical rotation photograph of ($H/R = \tan\mu$), C = crystal.

$R = \tan\mu$ (shown in Fig. 6.13), where μ is the layer line angle. The repeat distance $c = \lambda/\sin\mu$ where λ is the Miller's index $= 1$ for the first layer, 2 for the second layer, and so on. Since $l = 0$ on the 0 layer, the spots on the 0 layer is $hk0$. Similarly, the indices of the spots on the first layer will be $hk1$, on the second layer $hk2$, and so on. In general, one can directly measure the unit cell parameters by mounting the a or b axes as well. The rotation photographs can be interpreted using the reciprocal lattice concept. The indexing of this spot in the rotation photograph can be done with the use of what is known as Bernal's Chart. The principle and construction of this, again, is not within the scope of this book. Detailed treatments are given in the text books on X-ray crystallography.

6.6.3 Oscillation Method

The left- and right-handed sides of any rotation photograph are identical. This is because each reciprocal lattice point passes through the sphere of reflections in an identical manner on both sides of the direct beam. Similarly, the reciprocal lattice points are identical on the top and bottom side (i.e. $l = 1$ or $l = -1$). Thus, rotation photographs will have an inherent mirror symmetry. This can be partly circumvented (i.e., from right to left, mirror symmetry can be removed) if the crystal is rotated by a restricted angle, i.e. if it is oscillated back and forth through a limited angular range. This removes the artifact of symmetry in the rotation photographs. Therefore, the actual mirror symmetry, if present, can thus be discarded. Except for this advantage, the other features of the oscillation methods are the same as in the rotation method. From the oscillation photograph we can determine the unit cell dimensions and the reflection symmetry parallel to the oscillation axis. Though the indexing of oscillation and rotation photographs can be carried out, it is not unambiguous. This is because it is not known at what stage of oscillation, the reflection sphere passes through the reciprocal point. In other words, at what inclination of the incident X-ray beam the relevant lattice planes are so oriented as to satisfy Bragg's equation. This leads to the overlapping of some reflection spots. This can be resolved by moving the film concurrent to the oscillation or rotation of the crystal.

6.7 THE MOVING FILM METHODS

There are two methods by which the film can be moved: (1) the Weissenberg method and (2) the Precession method.

6.7.1   The Weissenberg Method

The arrangement in the Weissenberg camera is schematically shown in Fig. 6.14. As the crystal is oscillated by 180°, the film moves back and forth along the axis of rotation. These two movements are synchronized in such a way that the back and forth movement of the film and the clockwise–anticlockwise rotation of the crystal are exactly matched in space and time, i.e., for every one degree rotation of the crystal, the film cassette moves by 0.5 mm. Immediately inside the film cassette there is a cylindrical metal screen (opaque to X-rays) which is in two halves. The position of the gap between the halves of the screen can be adjusted in such a way that the diffracted beam from only one layer will reach the film in a given experiment. Figure 6.15 shows the sequence in which the diffraction spots are registered in a Weissenberg Camera. In this figure, only the central reciprocal lattice row or the trace is considered at a given time 't_1'. This row is at right angles to the direct beam. As the crystal is rotated anticlockwise, at another time 't_2', the central lattice row (OP) is rotated by an angle ω (omega). Then the diffracted beam emerges from the Ewald's sphere at an angle γ(gama) = 2ω. During this time, the film can be moved to a new position and the diffracted spot appears at P_1. If, on continued rotation of the crystal, the OP also rotates further, it touches the sphere at P_2 at time 't_3'. A new diffraction spot P_2 is registered. Further rotation produces a second line of spots on the photograph in the opposite direction away from the central beam position corresponding to P_3, P_4, etc. after time't_4, t_5', etc. This is because the

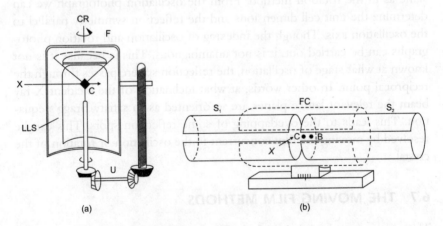

(a)   (b)

Fig. 6.14 The Weissenberg camera arrangement: (a) vertical (b) horizontal movement of layer screen. LLS = layer line screen, X = X-ray, CR = crystal rotation axis, U = oscillating spindle, FC = film cassette B = beam trap, S_1 and S_2 = split screens, C = crystal.

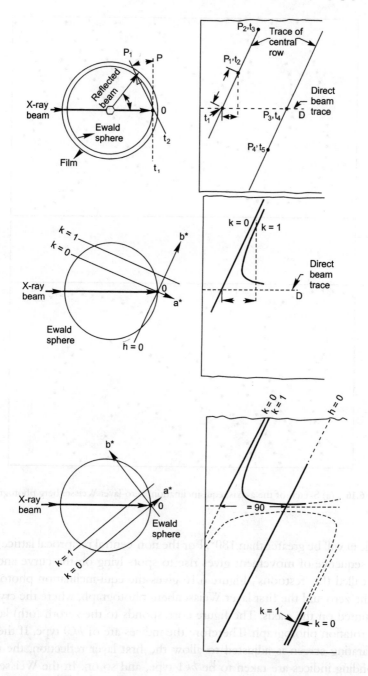

Fig. 6.15 Different positions of the reciprocal lattice with respect to the Ewald sphere and the sequence of appearance of the festoons in a Weissenberg photograph.

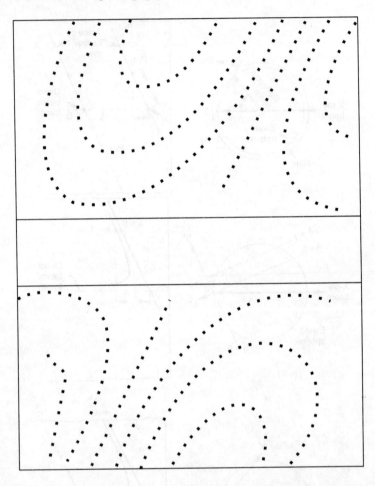

Fig. 6.16　(a) Sketch of the a-axis, equi-inclination, zero-layer Weissenberg photograph of CuS.

angle ω will be greater than 180°. For the non-central reciprocal lattice rows, the sequence of movement gives rise to spots lying on the curve and they are called the 'festoons'. Figure 6.16 gives the equi-inclination photograph of the zero and the first layer Weissenberg photograph, where the crystal is mounted on the c-axis. The figure corresponds to the zeroth (oth) layer in the rotation photograph. Therefore the indices are of $hk0$ type. If the layer separating screen is adjusted to allow the first layer reflection, the corresponding indices are taken to be $hk1$ type, and so on. In the Weissenberg photograph the indexing is unambiguous, i.e. $hk0$ and $\overline{hk}0$ can be distinguished. The indexing of the Weissenberg photograph can be done with the help of the Weissenberg chart.

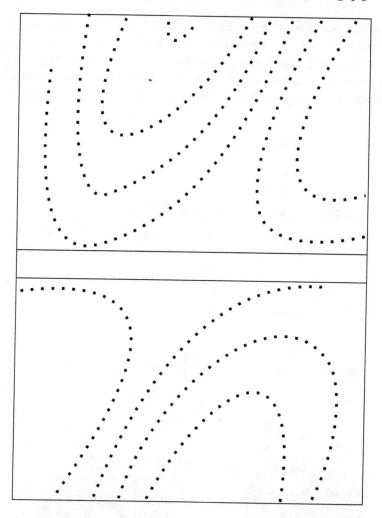

Fig. 6.16 (b) Sketch of the a-axis, equi-inclination, first-layer Weissenberg photograph of CuS.

6.7.2 Buerger's Precession Method

In the Weissenberg photograph, the reflections other than those from the central reciprocal axis ($h00$, $0k0$, $00l$ type) will fall on the curved lines or festoons. It fails to directly exhibit the symmetry of that crystallographic axis. The indexing is also more tedious. These features arise from the fact that the reciprocal row of lattice points are always tangential to the reflecting sphere, rather than moving along the curved surface of the reflection sphere as the crystal rotates. Then, these difficulties can be overcome if one adopts different types of movements for both the crystal and the film, i.e. they are made to precess. In the precession methods, the movement is the same as that of a rotating top. The experimental set-up is shown in Fig. 6.17. The precession of the crystal and the film are linked so that each moves identically along the direction of the incident beam. There is a screen attached to

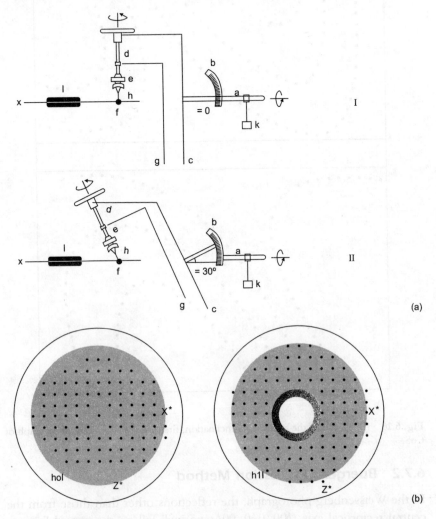

(a)

(b)

Fig. 6.17   (a) Precession camera mechanism sketch. a = motor driver spindle, b = graduated arc, c = film loader, d = goniometer spindle, e = goniometer, f = crystal, g = screen with slots, h = beam back-stop absorber, i = collimator. Diagram I indicates the arrangement where the precession angle μ = 0. In diagram II, μ = 30° (b) precession photograph on two different layers of a crystal.

the spindle to get the layer separation. The precession photographs are the direct representation of the reciprocal lattice net. The actual symmetry is directly observable as in the case of Laue photographs. The symmetry of the net is also directly visible and can be indexed without ambiguity.

6.8 SYSTEMATIC ABSENCES

When all the X-ray reflections are indexed using the above techniques, it will be noticed that one or the other types of *hkl* reflections are absent. These absences arise from the type of Bravais lattice or because of the presence of the screw axis or glide planes. Such systematic absences are important in fixing the space group symmetry.

The absence of reflections arising out of the Bravais lattice have been briefly mentioned in section 6.5.3, dealing with the indexing of cubic pow-der patterns. If a plane passes through a given lattice point, the translation symmetry requires identical and parallel planes to pass through each of the remaining similar lattice points. This imposes restrictions on the existing valid *hkl* planes, depending upon the type of Bravais lattice. For example, if a lattice point exists at the centre of the 001 face of a *c*-centred orthorhom-bic lattice, an additional plane (*hkl*) has to pass through this additional point (as compared to the corresponding primitive lattice). This is illustrated in Fig. 6.18a, which is the plane-view of an orthorhombic primitive lattice from the *c*-axis. The (210) planes are shown to pass through the lattice points. If the same lattice is *c*-centered (Fig. 6.18b), the (210) plane does not exist because, on an extended drawing, it does not encounter the identi-point with long range periodicity. On the other hand, the (420) type of reflections satisfy the long-range periodicity. If we consider the (310) plane on the *c*-centered lattice, it satisfies the above conditions and has a definite condition. One may note that (3 + 1) is an even number, so also the 420, 510 reflections. Whereas, in the 210 reflections, (2 + 1) is not even. This has no existence in a *c*-centered lattice. In general, (*h* + *k*) should be even in the *c*-centered lattice. Likewise, for *a*-face centered orthorhombic lattice, the *hkl* should follow (*k* + *l*) = even. In the *b*-face centered lattice (*h* + *l*) should be even. Table 6.7 gives the Bravais lattice-imposed restriction for the indices of the X-ray reflections.

The restrictions can also arise out of the presence of a screw axis or glide planes, because they also introduce additional points and hence additional planes. The detailed discussions on these are avoided for brevity, although the conditions are tabulated in Table 6.7.

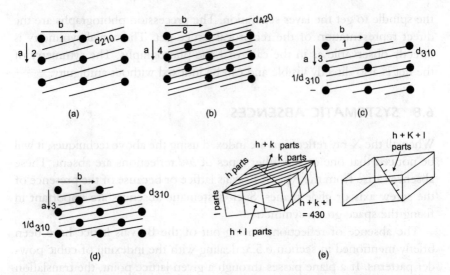

Fig. 6.18 Explanation to account for systematic absences (a) view down the c-axis of a direct primitive orthorhombic lattice to show d_{210} interplanar spacing (b) view of c-face centered orthorhombic lattice showing d_{420} lattice spacings; 420 reflection is valid in this case instead of 210 reflection (c) and (d) the corresponding lattices in a and b showing d_{310} planes indicating that 310 reflection is valid in both the cases (e) a set of translationally equivalent hkl planes in primitive direct lattices (see text).

From the restrictions of the X-ray reflections, one can deduce the lattice types and also the presence and types of screw axis or glide planes. This permits the identification of crystal space groups and, in most cases, it helps

Table 6.7   Systematic absences—general conditions

(a)   *Lattice-imposed conditions for Bragg reflections*

Lattice type	Conditions for possible reflection
P, R	None
A	$k + l = 2n$
B	$h + l = 2n$
C	$h + k = 2n$
F	$\left.\begin{array}{l} k + l = 2n \\ h + l = 2n \\ h + k = 2n \end{array}\right\} = h, k, l$ all odd or all even
I	$h + k + l = 2n$
R (If indexed on hexagonal axes)	$\left.\begin{array}{c} -h + k + l = 3n \\ \text{or} \\ h - k + l = 3n \end{array}\right\}$

(b) *Conditions for reflection imposed by screw axes*

	Screw Axis		Conditions for possible reflection
Type	Orientation	Translation*	
	Nonhexagonal Crystals		
$2_1, 4_2$	[100]	$a/2$	For $h00 : h = 2n$
$2_1, 4_2$	[010]	$b/2$	For $0k0 : k = 2n$
$2_1, 4_2$	[001]	$c/2$	For $00l : l = 2n$
2_1	[110]	$a/2 + b/2$	For $hh0 : h = 2n$
$4_1, 4_2$	[100]	$a/4$	For $h00 : h = 4n$
$4_1, 4_2$	[010]	$b/4$	For $0k0 : k = 4n$
$4_1, 4_2$	[001]	$c/4$	For $00l : l = 4n$
	Hexagonal Crystals		
6_2	[0001]	$c/2$	For $000l : l = 2n$
$3_1, 3_2, 6_2, 6_4$	[0001]	$c/3$	For $000l : l = 3n$
$6_1, 6_2$	[0001]	$c/6$	For $000l : l = 6n$

* The "o" subscripts are omitted from a_o, b_o, and c_o in this column, this being a common practice.

(c) *Conditions imposed on reflection indices by glide planes*

Glide plane			Permissible reflection indices**
Orientation	Symbol	Translation*	
	Axial glides		
[100]	b	$b/2$	For $0kl, k = 2n$
[100]	c	$c/2$	For $0kl, l = 2n$
[010]	a	$a/2$	For $h0l, h = 2n$
[010]	c	$c/2$	For $h0l, l = 2n$
[001]	a	$a/2$	For $hk0, h = 2n$
[001]	b	$b/2$	For $hk0, k = 2n$
[1$\bar{1}$0]	c	$c/2$	For $hhl, l = 2n$
[1$\bar{1}$00]	c	$c/2$	For $hh\bar{2}\,\bar{h}\,l, l = 2n$
[11$\bar{2}$0]	c	$c/2$	For $hh0l, l = 2n$
	Diagonal glides		
[100]	n	$b/2 + c/2$	For $0kl, k + l = 2n$
[010]	n	$a/2 + c/2$	For $h0l, h + l = 2n$
[001]	n	$a/2 + b/2$	For $hk0, h + k = 2n$
[1$\bar{1}$0]	n	$a/2 + b/2 + c/2$	For $hhl, 2h + l = 2n$
	Diamond glides		
[100]	d	$b/4 + c/4$	For $0kl, k + l = 4n$
[010]	d	$a/4 + c/4$	For $h0l, h + l = 4n$
[001]	d	$a/4 + b/4$	For $hk0, h + k = 4n$
[1$\bar{1}$0]	d	$a/4 + b/4 + c/4$	For $hhl, 2h + l = 4n$

* "a" subscripts omitted.
** The reflectional indices cited in this column represent sets of lattice planes that (a) are perpendicular to the glide plane and thus (b) are doubled or quadrupled in number by the glide-plane translation.

to narrow down the identification to a limited number of related space groups. The space groups that impose identical restrictions on h, k and l values cannot be distinguished by X-ray technique alone. They are grouped together under the name 'diffraction symbols'. This can be illustrated with an example of a tetragonal crystal, where all the reflections obey the conditions $h + k + l = 2n$ (even), meaning it is a body-centered lattice type with the symbol I. It is also observed that hhl has $2h + l = 4n$. This corresponds to a glide plane of the type d-glide. Therefore, the diffraction symbol I . . d. There are two space groups close to this diffraction symbol, namely, $I4_1nd$ and $I4_2d$. In order to distinguish the two, one has to resort to the morphological observations to detect the presence or absence of the mirror plane.

6.9 INTENSITIES OF DIFFRACTED X-RAYS AND STRUCTURAL ANALYSIS

This simplified section is intended for readers who are keen to know about the elementary principles of X-ray crystal structure analysis which essentially makes use of the intensities of reflected X-rays. The section, however, is not an in-depth treatment for those who want to use the diffraction technique as a tool in their scientific work. For this purpose, advanced textbooks in X-ray crystallography may be referred to.

6.9.1 Measurement of Intensities

Two methods are employed for the measurement of intensities of diffracted X-rays: (a) photographic (b) the ionization detector or 'scintillation counter' or solid state (semiconductor) detector. In the photographic method, the diffracted X-rays falling on the photographic plate darkens the film. The density of darkening is

$$D = \log I_0 / I, \qquad (6.12)$$

where I_0 is the intensity of incident light and I is the light transmitted through the film. The human eye can see the darkening of only up to two orders, i.e. $I_0 = 100$ and $I = 1$. The X-ray films are coated on both the surfaces and here one can detect darkening up to three orders. This darkening is quantified using an optical densitometer, wherein the standard films darkened with known exposures are used.

The use of the ionization detector or scintillation counters is well-known in the field of nuclear physics and radioactivity. The electromagnetic radiations with shorter wavelengths, (such as X-rays or γ-rays as also particulate radiations such as α and β-rays) can ionize a collection of gas molecules in a chamber and produce an electrical current proportional to the intensity of radiation. These radiations can also produce secondary, longer wavelengths of visible light by some solids through the process called scintillation. The intensity of light emitted by scintillation is proportional to the intensity of the impinging high energy radiation. There are two ways of measurements in practice: 1) fixed time method; 2) fixed count method. In the former case, it is the number of pulses across an unit area for a known time. In the latter case, it is the time required for recording a specified number of pulses or counts. The photographic technique is used when X-ray cameras are employed for recording the diffraction patterns and are relatively simpler and cheaper. The ionization detectors/scintillation counters are in use with the powder and single crystal diffractometer. The single crystal diffractometers will normally have the specimen movement involving more than one axis. The ultimate purpose is to measure the intensity of as many diffracted beams as possible emitted around the crystal.

6.9.2 Intensity and its relation with the Atomic Arrangement

The relation between the intensity of diffracted X-rays and the arrangement of atoms in a crystal can be best illustrated by considering different layers of small and large atoms, as given in Fig. 6.19. At some angle θ, the X-rays scattered by the plane of large atoms will be in-phase. Similarly, the X-rays scattered by planes of smaller atoms will also be in-phase. But the path lengths between the two sets will be different and will be less than 1λ. This means the two sets of atomic planes do not scatter in-phase. This leads to the partial or complete destructive interference, so that the amplitude of diffracted waves is less than what it would have been if they were scattered totally in-phase by both the sets of atomic planes. Since the intensity is proportional to the square of the amplitude, the above situation leads to the decreased intensity of the diffracted beam. This principle was first used by W H Bragg and W L Bragg to solve the structure of NaCl crystals. Intensities were measured by them using three differently orientated crystals of NaCl, namely 100, 110 and 111. In the first two cases, the intensities of higher order reflections (200, 400, 600 or 220, 440, 660, etc.) were nearly the same and showed a slight uniform decreasing with $\sin\theta$ values. Whereas along the 111 orientation, the 111, 333, 555 reflections showed very low intensities as compared to the 222, 444 and 666 planes. They concluded that along the 111 direction, the alternating planes were made up of either only Na^+ or Cl^- ions. This is because the scattering capacity of the atoms of Na and Cl are different. Further, when this experiment was repeated with

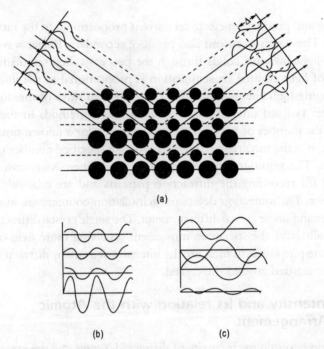

Fig. 6.19 (a) The relation of intensity of X-ray reflections to an arrangement of atoms. (b) Additive effect of waves in-phase. (c) Reduction of intensity in out-of-phase waves.

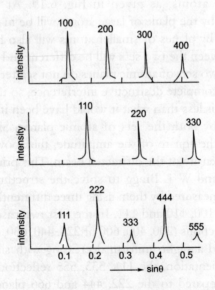

Fig. 6.20 Schematic illustration of intensities of reflection from first few orders of (100), (110) and (111) planes of NaCl crystal. The indices assigned to (100) and (110) reflections are incorrect (see text).

KCl (which is isostructural with NaCl), the 111, 333 and 555 reflections disappeared while the 222, 444, 666 reflections remained unaltered. This is because K^+ and Cl^- have the identical scattering power due to their identical electronic (isoelectronic) configuration. This destructive interference of the diffracted waves is complete in the case of KCl, whereas it was only partial in the case of NaCl. Thus, using the intensity relation, the crystal structure analysis was done wherein the NaCl structure was deduced to be consisting of two interpene-trating FCC lattices of Na^+ and Cl^- ions (See Chapter IV for halite-type structure). Accordingly, the indices shown on the upper two curves of Fig. 6.20 have to be either all even or all odd. This can be attained by doubling the indices, i.e., 200, 400, etc. as the first curve and 220, 440, etc. on the second curve.

Understanding the atomic arrangement is not always so simple as in the case of NaCl. The general relation between the intensity of diffracted X-rays and the atomic arrangement must be understood.

6.9.3 Atomic Scattering Factor

An electron coming in the path of the X-rays starts oscillating in response to the electrical vector of the X-rays, because an electron is a charged particle. The oscillating electron will start scattering the X-rays. This is a coherent scattering, i.e. it is not associated with any change in the energy or wavelength in contrast to the situations of X-ray fluorescence. When an atom comes in the path of the X-rays, each electron in the orbitals of the atom will scatter X-rays. This will be approximately Z times stronger than an individual electron, where Z = atomic number. One can define an electron scattering factor $f = E_a/E_e$, where E_a is amplitude of scattered X-rays by the atom and E_e is the amplitude of the scattered X-rays by an electron. The total scattering will decrease with the increasing scattering angle. This is because an atom has a finite volume and cannot be treated as a Euclidean point since the electrons form a charged cloud around the nucleus. The X-rays scattered by such moving electrons from different parts of the cloud, produce interference effects. Consequently, the X-ray-scattering is the sum of the contribution made from all the volume elements of the electron cloud in an atom. While considering this, one has to take into account the path differences between the scattered X-rays of each of the volume elements. This is done by what is known as a phase factor. A phase factor is a component of the atomic scattering factor, indicating the relation between the incident and the scattered beam to the position vector of each volume element. Considering that there is only one electron in an atom, these contributions were first calculated and then multiplied by the atomic number Z, thereby deriving the atomic scattering factor. These values are available for every element and their stable ions in the International Table of Crystallography. The intensity of X-ray diffraction is related to the square of the atomic scattering factor.

$$I_a = f^2 I_e \text{ or } I_a/I_e = f^2 \tag{6.13}$$

Hence, one measures the intensity and arrives at the scattering factor. The scattering factor f is dependent on the scattering angle θ and λ (i.e. $\sin\theta/\lambda$).

6.9.4   Scattering of X-rays by Molecular Aggregates

Similar to the interference of X-rays scattered by electrons from different volume elements of the electron cloud in an atom, it is also possible that the interference can take place between the X-rays scattered from different molecules when they exist in aggregates. If the molecules are polyatomic, the atoms are chemically bonded. The interference of scattered X-rays from each of the atom in a given molecule gives rise to the internal interference effects. It is also possible to have interference effects between scattered X-rays from different molecules when they exist in aggregates. This leads to external interference effects. The relative importance of these two effects, therefore, depends on the molecular complexity. The simplest situation is to consider a monoatomic gas where the external interference effect is the only dominant effect. In the monoatomic gas, the distance between the atoms as well as the positions occupied by the atoms are continuously changing. Therefore, one has to consider a situation at a given point of time so as to have a meaningful understanding of the distances and positions. In Fig. 6.21, one of the atoms is considered as the origin, so that the total intensity scattered by all the atoms corresponds to that with reference to the

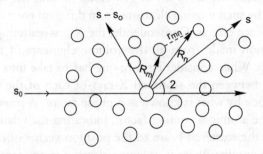

Fig. 6.21   Scattering of X-rays by monoatomic gas.

atom at the origin. Because the gas molecules are randomly moving, the origin atom itself is interchangeable. For the same reasons, the absolute distance of the individual atoms from the origin is also changing. As a result, only the average interatomic distances can be considered. This leads to an expression for the intensity of scattered X-rays in the form of the double summation,

$$I = \sum_m \sum_n f_m f_n e^{\left(\frac{2\pi i}{\lambda}\right)(S-S_0)\gamma_{mn}}, \tag{6.14}$$

where $(S - S_0)$ is called the diffraction vector and S_0 and S are vectors of incident and scattered beams: $\gamma_{mn} = R_m - R_n$, the interatomic distance between the two atoms m and n. If the time average intensity based on the above equation is plotted as a function of $\sin\theta/\lambda$, for a fixed value of γ_{mn} the curve will show stronger maximum towards low values of $\sin\theta/\lambda$ and weaker maxima at higher scattering angles (Fig. 6.22a). This demonstrates that even in a randomly arranged monoatomic gas there will be scattering in-phase at angles greater than zero degree although there is no orderliness in the molecular arrangement. Experimentally, it is difficult to demonstrate using X-rays, because of the broadening of the direct beam and the scattering efficiencies decrease proportionally to λ^3. Therefore, the wave length of the radiation has to be small in order to detect this 'in-phase' maxima. This is possible only by electron diffraction, where the λ is of the order of 0.01 to 0.1Å (whereas, $\lambda = 1.542$ Å for CuK_α X-rays).

Let us now consider the X-rays scattered by atoms of monoatomic liquids. The average interatomic distances in liquids are much smaller than in gases, as a result of denser packing. The existence of a short-range order is, therefore, inevitable in a liquid because the atoms cannot move far away from one another. At the same time, they cannot penetrate one another. The minimum distance they can be close, corresponds to twice their molecular radius. Under such a situation, imagine any given atom at the origin within the liquid. As we move farther from the origin, the number of atoms surrounding the origin at a constant radius (spherical shell) will keep increasing, although the atom per unit volume is constant. In such a spherical surface, $4\pi r^2(\rho r)$ represents a distribution function called the radial distribution function where $roe(\rho)$ is the density (i.e., no. of atoms per unit volume) at a distance r away from the atom at the origin.

The time average intensity given in Eq. 6.14 can be modified by multiplying it with the radial distribution function which will be applicable to liquids and vitreous substances. If this modified expression for the intensity is plotted as a function of $\sin\theta/\lambda$ (Fig. 6.22b), the intensity maximum shifts away

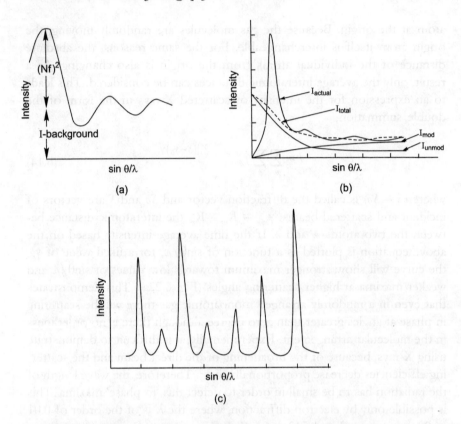

Fig. 6.22 X ray scattering intensity for: (a) monoatomic gas, (b) monoatomic liquid, and (c) monoatomic polycrystalline solids (schematic).

from the origin unlike, that for the gases. Furthermore, the peak heights are larger. Besides, more number of minor maxima appear in higher scattering angles. These features can be experimentally detected in X-ray diffraction of liquids.

In the crystalline solids, there is a long-range order because of the three-dimensional periodicity. Therefore, it is sufficient to calculate the intensity of X-rays scattered by atomic aggregates in an unit cell. This is so because all unit cells are identical. X-rays scattered by each unit cell are equivalent to the external interference effects. The position vector of the atoms in a unit cell and the origin of the unit cell are well-defined in a crystalline substance. The average interatomic distance γ_{mn} is replaced by a position vector R which is equal to $(ha + kb + lc)$, where a, b and c are unit cell parameters and h, k and l are the Miller indices.

The expression for intensity (I) in crystalline solids can now be expressed in terms of the triple summation for unit cells containing a single atom. For the unit cell containing n atoms, one more summation is necessary so that

$$I_n = \sum_n \sum_h \sum_k \sum_l f_n e^{\left(\frac{2\pi i}{\lambda}\right)(S-S_o)(b_a + k_b + l_c)} \tag{6.15}$$

It is possible to consider the intensity of X-rays scattered by one unit cell called the structure factor, F. The intensity is equal to F^2 or $F \times F^*$, where F^* is the complex conjugate of F. Therefore,

$$F_{(hkl)} = \sum_{n=1}^{N} f_n e^{\left(\frac{2\pi i}{\lambda}\right)\left(b_x + k_y + l_z\right)} \tag{6.16}$$

The structure factor F is better understood by considering a crystal with known atomic positions as examples. Let us consider a Cu crystal having a *FCC* lattice. The Cu atoms are at (000), (1/2, 1/2, 0), (1/2, 0, 1/2) and (0, 1/2, 1/2) for a given unit cell. These positions correspond to the unit cell origin or any one of the face-centres. The atoms at the other corners are not considered because they correspond to the origin of the adjoining unit cell. The structure factor for copper crystal is

$$F_{(hkl)} = f_{Cu}\left[e^0 + e^{\pi i(n+k)} + e^{\pi i(h+l)} + e^{\pi i(k+l)}\right] \tag{6.17}$$

which is equal to $4f_{Cu}$ when *hkl* are unmixed, and it is zero when *hkl* is mixed.

This means that the structure factor F is four times the atomic scattering factor of Cu at a given $\sin\theta/\lambda$ in these planes with unmixed *hkl* values. On the planes of all other Miller indices, the structure factor will be zero.

Let us now consider an NaCl crystal made up of two interpenetrating *FCC*, i.e. (individual sub-lattices are also *FCC*'s) of Na^+ and Cl^- ions. Here Na^+ ions are at (0, 0, 0), (1/2, 1/2, 0), (1/2, 0, 1/2), (0, 1/2, 1/2) and Cl^- ions at (1/2, 1/2, 1/2), (0, 0, 1/2), (1/2, 0, 0), (0, 1/2, 0) positions.

Therefore,
$$F_{(hkl)} = f_{Na^+} + \left[e^0 + e^{\pi i(h+l)} + e^{\pi i(h+l)} + e^{\pi i(k+l)}\right]$$
$$+ f_{Cl^-}\left[e^{\pi i(h+k+l)} + e^{\pi il} + e^{\pi ih} + e^{\pi ik}\right]$$
$$= 4\left(F_{Na^+} + f_{Cl^-}\right) \text{ where } hkl \text{ are all even}$$
$$= 4\left(F_{Na^+} - f_{Cl^-}\right) \text{ where } hkl \text{ are all odd}$$
$$= 0 \text{ when } hkl \text{ are mixed} \tag{6.18}$$

Since $F^2 = I$ (intensity), when $F = 0$, there will be no observable intensity of scattered X-rays arising from these planes. For example, when hkl is mixed, $I = 0$. Though the intensity will be observed when hkl are all even for NaCl, because the contribution from Na^+ and Cl^- are all additive. In comparison, the intensity will be lowered when hkl are all odd, because the corresponding structure factor arises from the difference between the atomic scattering factors of Na^+ and Cl^-. On the same grounds, for KCl crystals the intensity will be zero for all odd reflections, because $f_{K^+} = f_{Cl^-}$. It is also evident from the above example that if the atomic positions are known, the intensity of the X-ray reflection can be calculated. However, for unknown crystals the converse is followed, i.e. from the observed intensities of (hkl) reflections, the atomic positions are deduced through a series of calculations. The mathematics involved in such calculations is far more complicated and has to be referred to in well-known textbooks on X-ray crystallography, some of which are listed at the end of this book. Qualitatively, the procedure can be visualized as follows; the structure factor F is related to electron density ρ_{xyz} which is the sum of the electron density of individual atoms

$$F_{(hkl)} = \iiint \rho_{xyz} e^{\left[2\pi i \left(\frac{hx}{a} + \frac{ky}{b} + \frac{ly}{c}\right)\right]} V/abc\; dxdydz \qquad (6.19)$$

where V is the unit cell volume and V/abc is the geometrical factor to normalize the coefficients for non-orthogonal cells. Since the electron density and structure factor are well-behaved functions in a three-dimensional periodic lattice, they can be transposed by a Fourier series:

$$\rho_{(xyz)} = \frac{1}{V} \sum_{h}^{+\infty} \sum_{-\infty} \sum_{l} F_{hkl} e^{-2\pi i \left(\frac{hx}{a} + \frac{ky}{b} + \frac{lz}{c}\right)} \qquad (6.20)$$

It is not possible to calculate the three-dimensional electron density and sum it up all at once. Instead, electron density calculations for two-dimensional cross-sections are carried out at regularly spaced intervals along a given axis to arrive at the three-dimensional density. The electron density represents the distribution of atoms in a unit cell, because atoms of different elements have different electron densities as explained earlier. However, the above summations are not straightforward because experimentally we can only determine magnitudes of F; whereas, the individual phase factors remain unknown and will have to be determined separately. This, again, is a tedious task and an indirect determination is adopted. One of the methods

is to use a Patterson function which actually involves the conditional operations. Briefly, the structure factor is related to the Patterson function by the
expression:

$$P_{(xyz)} = \frac{1}{V^2} \sum_{k} \sum_{-\infty}^{+\infty} \sum_{l} F^2_{(hkl)} \cos 2\pi \left(\frac{hx}{a} + \frac{ky}{b} + \frac{lz}{c} \right) \qquad (6.21)$$

Since $F^2 = I$, and $F_{hkl}^2 = F_{hkl} \cdot F_{\overline{hkl}}$, the $P(xyz)$ can be experimentally
calculated for various cross-sections as mentioned above. A plot of $P(xyz)$
values for various cross-sections can be drawn as two-dimensional maps.
They are called Patterson maps. In these maps, the electron density peaks
recur in the same place more than once (superposition). If there are n peaks
in the electron density, then n peaks will be superimposed on the origin and
n^2-n peaks will be distributed at the rest of the unit cells. In this way the
atomic positions can be located in the unit cell, but need not necessarily
identify the different types of atoms. A typical example of the Patterson
map is shown in Fig. 6.23.

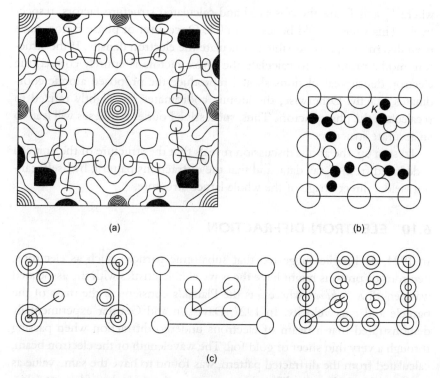

(a)　　　　　　　　　　　　　　　(b)

(c)

Fig. 6.23　The electron density map of KO_3 crystal on the (xy) plane: (a) the typical Patterson
function diagram, (b) corresponding projection of the crystal structure, (c) drawing the
Patterson diagram of the function at specific locations.

There are other routes commonly adopted for structural analysis in X-ray crystallography. For example, when a crystal contains heavy atoms (i.e., one with very many more electrons than the rest of the atoms), the structure factor can be determined unambiguously because the heavy atom can be precisely located in the unit cell. It is possible to calculate electron density at once using the measured F values and the calculated phase factors. Using these values the other atomic positions can be determined. An extension of this method is the isomorphous substitution method, wherein a light metal atom is replaced by a heavy atom. Then the heavy atom method is adopted. Image-seeking methods, anomalous scattering methods and the statistical methods using the theory of inequalities are the other direct methods. Whatever the method followed for X-ray structural analysis, the correction of the determined structure is checked by calculating the residual R, which is given by the expression

$$R = \sum \left[\left| F_o \right| - \left| F_c \right| \right] \Big/ \sum \left[\left| F_o \right| \right] \tag{6.22}$$

where F_o and F_c are the observed and calculated structure factors, respectively. This value should be less than 0.5. Once a set of phase factors have been determined, it is possible to calculate the electron density. If the structure model proposed to calculate the phase factors is entirely correct, the electron density calculations should reproduce the proposed atomic coordinates exactly. Otherwise, the atomic coordinates are slightly refixed to recalculate the phase factors. Thus, successive Fourier syntheses will lead to an accurate structure.

To sum up, the above discussion reveals that the structure of the unit cell is deduced from X-ray data, and that the diffractions pattern, by itself, cannot yield a direct image of the whole lattice structure.

6.10 ELECTRON DIFFRACTION

In 1924, de Broglie suggested that subatomic particles such as electrons, neutrons or protons might have the wave characteristics with the associated wavelength $\lambda = h/mv$, where h is the Planck's constant, m the mass of the particle and v its velocity. In 1927, Davisson and Germer experimentally demonstrated that a beam of electrons undergo diffraction when passing through a very thin sheet of gold foil. The wavelength of the electron beam, calculated from the diffracted pattern, was found to have the same value as predicted by de Broglie. The electrons emitted from a hot filament (electron gun) and maintained in a high vacuum are accelerated through a high potential difference of $> 10^5$ volts, so that the electrons acquire kinetic energy $= eV$, where e is the electronic charge and V is the applied potential. In fact, the velocities of these electrons are comparable to the speed of light so that a relativistic correction has to be made for the de Broglie relation. Thus, following this correction, the wavelength of the electron is expressed as

$$\lambda = h\left[2m_0 eV\left(1 + \frac{eV}{2m_0 c^2}\right)\right]^{1/2} \tag{6.23}$$

where m_0 is the rest-mass of electron and c is the speed of light. According to this relation, when the applied potential is 10^5V (100KV), λ will be about 0.037Å. Here, the wavelength of the electron beam can be varied with the applied potential, and the values will be around 1/100 of the normal wavelength of characteristic X-rays used for diffraction studies. Therefore, the diffraction pattern even at a very small scattering angle can be observed; for example, from that of monoatomic gases discussed in previous section.

Electrons are scattered by electric fields prevailing within the atom. Unlike the beams of the X-rays, the electron beam can be brought into focus either by electrostatic or electromagnetic fields so that the electron lenses which function analogous to optical lenses in a microscope can be produced. At the same time, the formation of a diffraction pattern is an inevitable step in the electron image formation. Thus, in a modern transmission electron microscope (TEM) both magnified images of the object as well as

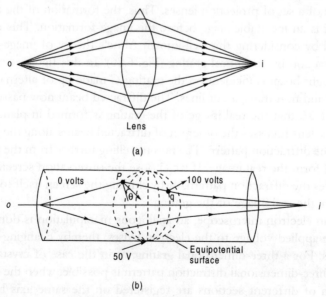

Fig. 6.24   (a) The comparison of a glass lens with (b) an electrostatic lens.

the electron diffraction patterns can be produced. Since TEM makes use of electrostatic lenses, let us see how an electrostatic lens works. As shown in Fig. 6.24, the force exerted on the unit charge (electron) is always at right angles to the equipotential surfaces. At '*p*', the electron is deflected towards the axis; whereas, at '*q*' it is deflected away from the axis. These two deflections from *p* and *q* do not cancel each other completely since the right angle component of the force at the left half of the sphere is larger than that in the right half. Therefore, the electron beam converges onto a point. Further, by applying different values of the positive potential on the right-hand hollow cylinder, the focal point of convergence of the electron beam can be varied. This is analogous to the focussing by a compound optical lens, through the mechanical movement.

The TEM contains a set of electron lenses whose layout is shown in Fig. 6.25. The condenser lenses focus the electron beam onto the object, which should be sufficiently thin to enable transmission of the electron beam. The objective lens generates the first stage diffraction pattern out of the scattered electron beam. The electrons spread out from the diffraction spots to cover the area of the image-forming plane and at this image-forming plane, the intermediate lenses are located. The image formed at the intermediate plane acts as though it is the original object. The sequential process of the formation of a diffraction pattern followed by the image, repeats at the projection lens. This helps in the formation of a highly magnified version of the original object as well as of its diffraction pattern through the use of a set of projection lenses. Thus, the formation of the diffraction pattern is an inevitable stage before the image-formation. This can be explained by considering the well-known Abbe's theory of image-formation in diffraction by an optical grating (Fig. 6.26). In this figure, a monochromatic light beam is incident on the grating *G* made up of alternating transparent and non-transparent lines. The diffracted beam now passes through a lens *L* so that the real image of the grating is formed in plane *I*. Before that, the lens focusses the other set of diffracted beams along the plane *D* to form the diffraction pattern. The rays travelling further from the plane *D* to plane *I* form the real image. If we choose the observation screen at *D*, one observes the diffraction pattern, or by shifting the screen itself to *I*, one can observe the real image of the grating.

In an electron microscope, an analogous manipulation is done by varying the applied voltage to the electron lenses, thereby changing their focal lengths. For a three-dimensional grating, as in the case of crystals, a complete three-dimensional diffraction pattern is possible, when the diffraction pattern of different sections are registered on the same axis by electron

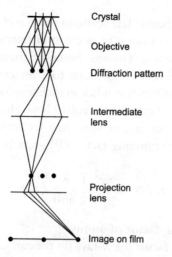

Crystal

Objective

Diffraction pattern

Intermediate lens

Projection lens

Image on film

Fig. 6.25 Configuration of electron lenses and image-formation in an electron microscope.

diffraction. If a single beam is used to form the image, a bright field micrograph is obtained. The image, produced by a set of diffraction beams from a set of (hkl) planes, produces (avoiding direct beam) the dark field image. The direct beam can be avoided by suitably tilting the specimen.

From the above discussion, it is clear that very thin specimens have to be used for electron diffraction (about a few hundred Å). Another limitation of the electron microscope is that the orientation can be varied only by about ±30° in relation to the direction of the incident beam. Furthermore, the crystal should be able to withstand a high vacuum and the local heating produced by the electron beam. However, the heating will be minimized by increasing the accelerating voltage, thereby decreasing the effective wave-

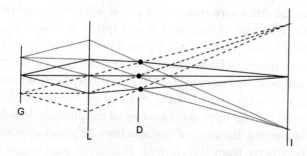

Fig. 6.26 Abbe's theory of image-formation in diffraction by an optical grating.

length of the electron beam. The scattering with a change of energy (inelastic scattering) is more common in electron diffraction. This, many a time, complicates the diffraction patterns. Some of these disadvantages do restrict a electron diffraction as the basic tool for crystal structure analysis. Above all, the electron diffraction is not an exclusive tool for structure analysis, because the scattering of electrons involves both electrons and the nuclei of the atoms. Generally, the atomic scattering factor for an electron is compared with the atomic scattering factor of X-rays by the relation

$$f_e / R = \frac{me^2}{2h^2 R} \left(\frac{\lambda}{\sin\theta} \right) (Z - f_x) \tag{6.24}$$

f_e = electron scattering factor of atoms
f_x = atomic scattering factor for X-rays (in the earlier sections denoted as f)
R = distance from the atoms at which the amplitude is measured
Z = atomic number

A further complication is that f_e does not continuously decrease with $\sin\theta/\lambda$ nor is it proportional to Z, the atomic number. For lighter elements f_e decreases with Z at small angles. On the average at $(\sin\theta/\lambda)$ around zero, f_e is proportional to $3\sqrt{Z}$ so that lighter elements scatter electrons better than the X-rays.

6.10.1 Types of Electron Diffraction Patterns

If the wavelength of the electron beam is of the order of 0.05Å, then the radius of Ewald's reflection sphere will be 20 Å⁻¹. In comparison, the reciprocal lattice constants will be of the order of 0.2 to 0.4 Å⁻¹. For a single orientation of a crystal, the number of reciprocal points intersecting Ewald's sphere is limited (see Fig. 6.27). Here, the electron beam is striking along the [001] direction, perpendicular to the [0*k*0] direction of an orthorhombic crystal. Then the diffraction spots at the centre will correspond to scattering from the 0*k*0 reflection, where the Ewald's sphere is intersecting the 0*k*0-type reciprocal lattice points. The reciprocal points from the *hk*1 plane intersect Ewald's sphere in limited numbers. Therefore, in the diffraction pattern, there will be a gap between the *hk*0 and *hk*1-type reflections, the latter appearing as concentric rings. Usually, most of the electron diffraction patterns show only the central portion because of the difficulty in achieving the ideal orientation and thinness of the specimen. The measurement of the reciprocal parameter from the electron diffraction pattern can be understood from Fig. 6.28. The arc *OQ* represents a part of the reflection sphere, since the wavelength of the electron beam is very small compared to the reciprocal lattice distances. The arc distance OQ can be equated to the straight line distance, OP. This is all the more valid because the Bragg angles can be as low as $1°$ in electron diffraction. Hence, in Fig. 6.28, L becomes the effective distance between the object and the film. The reflection spots are spaced at a distance R.

Therefore,

$$\tan\theta = R/L \qquad (6.25)$$

Because the value of θ is small, and $OP = OQ = d^*$ and

$$OQ = 2\sin\theta/\lambda = 2\theta/\lambda \qquad (6.26)$$

Therefore,

$$d^* = 2\theta/\lambda = R/L\lambda \quad \text{or} \quad d = \lambda L/R \qquad (6.27)$$

Experimentally, neither λ or L is determined separately; λL is determined together as a camera constant using a standard material of a known reciprocal lattice constant (gold, mica, etc.). Using the camera constant, the reciprocal lattice of the known material is determined.

Commonly, three types of electron diffraction patterns are observed.

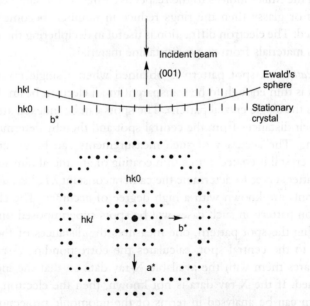

Fig. 6.27 Formation of diffraction pattern by a monochromatic electron beam incident on a crystal, parallel to the (001) direction.

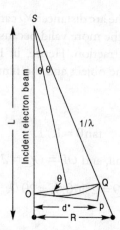

Fig. 6.28   The generated electron diffraction pattern showing the relation between reciprocal parameter d*, λ and θ, as explained in the text.

1. *Ring Pattern*: This consists of a series of concentric rings observed for
   polycrystalline samples. From the camera constant and the distance
   between the rings and the central spot, one can calculate the reciprocal
   lattice constants. If the corresponding X-rays data is available, one can
   assign the Miller indices to the respective rings. If the samples are amorphous or glassy then the rings reduce in number, become broad and
   diffused. The electron diffraction is useful in deciphering the truly amorphous materials from cryptocrystalline materials.

2. *Spot Pattern*: A spot pattern is obtained when a single crystal is used,
   which is thin enough to be transparent to an electron beam. Figure 6.29
   shows the single crystal pattern. The spots can be indexed after measuring their distances from the central spot and thereby determining the *d*-
   spacing. The accuracy of such measurements can be enhanced if the
   single crystal is coated with a thin coating of any metal film such as gold.
   The latter serves to determine the camera constant λL, because its lattice
   constants are known with a high degree of accuracy. The electron diffraction pattern in such cases will be spots superimposed on rings. For
   indexing the spot pattern, one measures the distances of the 3 nearest
   spots to the central spot, calculates the corresponding *d*-spacings and
   compares them with the available X-ray data so that the index can be
   assigned. If the X-ray data is not known, then the electron diffraction
   pattern can be analysed in terms of the gnomonic projections of reciprocal lattices. The principle of gnomonic projections are explained in
   an earlier chapter. The difference here is that we consider the planes of

reciprocal lattices instead of the morphological planes of the crystal. More details on this may be had from books on electron diffraction.

3. *The Kikuchi pattern*: If the specimen is reasonably thick, i.e. about half the maximum penetration distance from the electron beam, the Kikuchi patterns can be observed. This consists of pairs of parallel bright and dark lines. They arise from the inelastic scattering of electrons by the thicker specimens. The Bragg reflection of inelastically scattered electrons are possible, and the Bragg condition changes with the direction of propagation of these electrons. As they emerge out of the specimen, they interfere with one another to produce bright and dark lines (Fig. 6.30).

In summary, it can be said that electron diffraction studies have become

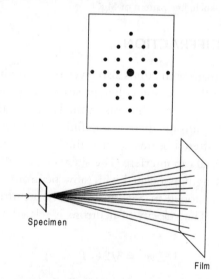

Fig. 6.29 Electron diffraction spot pattern obtained for a thin crystallite.

a strong tool for the investigation of nanometer-sized or cryptocrystalline materials. It can distinguish truly glass or amorphous material from the cryptocrystalline samples. Both these types of solids will show broad X-ray diffraction patterns. It is also a good tool in the study of defects such as line defects, because the electron scattering and diffraction conditions will be different in and around the immediate vicinity of line-defects. The extended lattice defects and intergrowths can be studied by the electron imaging technique, which is an extension of electron microscopy.

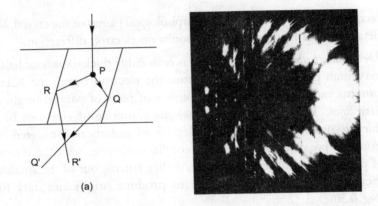

Fig. 6.30   (a) The principle of formation of Kikuchi lines (b) Combined electron diffraction spots and the Kikuchi line pattern of MgO.

6.11   NEUTRON DIFFRACTION

The interaction of neutrons with atoms also gives rise to both coherent and incoherent scattering. Of these, the coherent scattering gives information about the arrangement of the constituents in the lattice. In the case of the coherent scattering of neutrons, there is a definite phase relation between the incident and the scattered beams, so that the beam scattered from different atoms in the crystal can interfere. The situation is analogous to X-ray diffraction. The wavelength associated with slow neutrons is given by the de Broglie relation mentioned in Section 6.10. The neutrons are tapped from a nuclear reactor. The velocity of neutrons is related to the temperature of the reactors:

$$1/2mv^2 = 3/2k_BT \qquad (6.28)$$

where v is the Root Mean Square Velocity (RMS) of neutrons and k_B is the Boltzman constant. In practice, the temperature of the nuclear reactor is such that $\lambda = 1.32$ to 1.6 Å which is appropriate for diffraction from crystalline lattices because of the comparability (in wavelength) to characteristic X-rays. It is evident, therefore, that neutron diffraction cannot be carried out in all laboratories except those connected with atomic reactors.

The neutron beam tapped from the reactor is not monochromatic. Therefore, the collimated neutron beam is made to incident on a single crystal (graphite, quartz, etc.) so as to satisfy Bragg's relation and thereby obtaining a monochromatic beam at a definite angle. However, the neutron beam from the reactor in any direction has a lower intensity than that of an X-ray beam from a normal X-ray unit. Therefore, a large cross-sectional area of both the single crystal monochromator as well as the specimen has to be used, in order to produce adequately measurable intensity. Thus, the large specimen size in neutron diffraction studies is a prerequisite. Although powder diffraction is not effective, it definitely precludes single crystal studies on those materials which do not form large single crystals with an area of greater than a few square centimetres.

6.11.1 Mechanism of Scattering of Neutrons

The mechanism of scattering of neutrons differs, depending upon whether the atomic species is paramagnetic or otherwise. In diamagnetic solids, the neutrons are scattered solely by the nucleus of the atoms because the electrons being too small cannot deflect the neutron beam. When a plane wave of even amplitude is incident on the nucleus of an atom in a solid, the nucleus is not free to recoil. Under such conditions, the amplitude of scattered neutrons is given by b/R, where R is the distance from the scattering centre at which the amplitude is measured and b is the scattering length. Both R and b are equivalent to the position vector and the atomic scattering factor, respectively, in the X-ray scattering. This means that the scattering length is a nuclear property and is different for nucleii with different sizes. Unlike the atomic scattering factor (f), b is nearly independent of the scattering angle. This is because the radius of the nucleus is very small compared to the neutron wavelength. It is also because f arises from the scattering of X-rays from the electron cloud, and the size of the electron cloud is comparable to the X-ray wave length. Another difference is that b is independent of the wavelength of the neutron. The relation between b and the atomic species is much more complicated than that of f and the atomic species. Nucleii of different isotopes of the same element differ in b, so that one can talk only about the average scattering length $\bar{b}^*$ which is equal to $b \times$ isotopic abundance ratio for various isotopes. If the isotope has non-zero nuclear spin moments, its scattering length has two possible values and must be taken into account in calculating the coherent scattering distance $\bar{b}^*$. Because of these complications, a theoretical prediction of $\bar{b}^*$ has been impossible. On the other hand, values of $\bar{b}^*$ have to be experimentally determined.

The inter-relation between $\bar{b}^*$ and f can be given by

$$\bar{b}^* = e^2 f / mc^2 \tag{6.28}$$

showing that $\bar{b}^*$ is roughly of the same magnitude as f, so that crystals scatter neutrons as well as electrons by nearly the same amounts. $\bar{b}^*$ also varies irregularly with the atomic number. For a few elements, such as ^{1}H, ^{48}Ti, ^{62}Ni, the scattering length is negative. The sign of b corresponds to the change in the phase difference during scattering. The phase difference is π for positive b and zero for negative b.

Neutrons interact with the permanent magnetic moment of atoms or molecules containing the unpaired electrons. This is because, although the neutron is electrically neutral, it has finite magnetic moment. The neutron–electron (unpaired) interaction is such that the amplitude of the scattered neutrons falls off with increasing $\sin\theta/\lambda$, which is analogous to X-ray scattering. For the ferromagnetic, antiferromagnetic and ferrimagnetic substances, the neutron scattering is coherent. In comparison, the neutron scattering in paramagnetic substances is incoherent. This contributes to the background scattering. Thus, neutron diffraction is used to study the alignment of magnetic moments in solids, particularly polarized neutrons, i.e. neutrons whose spins are aligned in one direction. Through X-ray diffraction, one gets information about the atomic arrangement in a lattice. The neutron diffraction gives us additional information on the magnetic alignment or what is generally called the magnetic structure of solids. In magnetically-ordered materials, the neutron diffraction pattern is considerably modified in comparison to the X-ray diffraction pattern.

Let us take the example of the *BCC* lattice [Fig. 6.31a]. As explained in Section 6.5.3, the X-ray reflection appears in the order of increasing 2θ values: 110, 200, 211, 222, etc. reflections (i.e. $h + k + l$, all even). If the same lattice is occupied by a magnetic species, say MnF_2, then the magnetic moments on these molecules are antiferromagnetically arranged [Fig. 6.31b], although the crystal packing is the same as in the normal *BCC*. The neighbouring atoms are anti-parallel and the substance as a whole does not possess a net magnetic moment. The neutron diffraction will now show reflections as though the lattice is primitive, i.e. reflections such as 100, 110, 111, 200, 210 , etc.($h + k + l$ = no restrictions), some of which are not allowed in the *BCC*. This is because the differently oriented magnetic moments can be distinguished by neutron diffraction. The above difference in the diffraction pattern arises because of the fact that there are two possible ways of arranging the spin, i.e. spin up and spin down in the same lattice. Instead of the spin change an additional symmetry operation in terms of colour change (say from black to white or green to blue and so on), can be considered. In this way, the conventional crystallography which centres around the three symmetry operation (i.e. around a point, a line or a plane) will have another symmetry operation (colour symmetry) to be added on. This will lead to additional point groups, Bravais lattices and space groups,

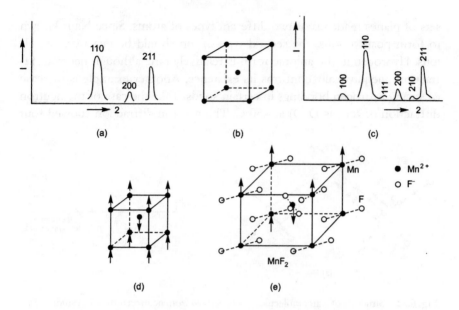

Fig. 6.31 Neutron diffraction patterns of: (a) The non-magnetic *BCC* structure shown in the figure (b) and of the antiferromagnetic structure shown in (d). Note that the diffraction pattern in (c) is apparently that of a primitive cubic structure (e) the crystal structure of the antiferromagnetic MnF_2 wherein every lattice point is occupied by the F-Mn-F molecule.

i.e. when the colour symmetry operation is included, the 32 point group becomes 122, 14 Bravais lattices become 36, and 230 space groups become 1651. The concepts of colour symmetry are developed in the famous book by Shubnikov and Belov.

One of the limitations of X-ray crystallography is in locating the light elements, particularly hydrogen; whereas, in neutron scattering, hydrogen can be located with better precision because the isotope ^1H has a negative scattering length. The meaning of the negative sign has been explained earlier. This has great consequences in studying organic crystals and is particularly of biological interest. Therefore, the neutron diffraction is often used in studying the crystals of organic compounds with complicated structures. The location of hydrogen is illustrated with a simple example of NaH. X-ray diffraction showed a Cu-type *FCC* pattern, implying as though H does not exist in the NaH crystal lattice. At the same time, metallic Na has a *BCC* lattice and cannot account for the Cu-type lattice of NaH. This problem was solved by neutron diffraction which showed a NaCl-type X-ray pattern, i.e., 111, 333, 555 reflections have low intensities as compared to 222, 444 reflections (see section 6.9.2). This implies that there should be two sets of planes made up of two different types of atoms. Since Na is known to form positive ions, the second type of ion should be negative, i.e., H⁻ ions. Hydrogen in the anionic form is relatively rare, although neutron diffraction unequivocally confirms its existence. Another example is the case of strong hydrogen bondings in certain solids. Take the case of the neutron diffraction of Ice (as D_2O) at –50°C. The neutron diffraction showed four

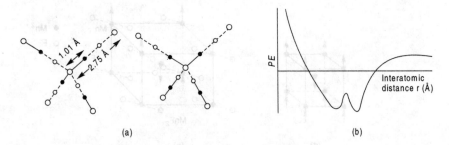

(a)   (b)

Fig. 6.32   Structure of water molecules in ice, derived from neutron diffraction studies. Two of the six possible configurations of hydrogen atoms around an oxygen atom are shown. Large open circles represent oxygen atoms; small circles represent hydrogen sites, filled circles, when occupied and open circles when vacant. Full lines correspond to covalent bonds and dashed lines are hydrogen bonds. (b) Double minima in potential energy for a hydrogen atom involved in hydrogen bonding.

half-atoms of deuterium around each oxygen atom at a distance of 1.01 Å, and also four half-atoms of deuterium at a distance of 1.75 Å. Totally, there is one deuterium between two oxygen atoms (Fig. 6.32a). This situation is also found prevailing in normal H_2OC (ice) at –50°C. Since the atoms cannot be split into half, there should be a new interpretation of the above observation. This leads to the concept of the double potential energy minimum in a hydrogen-bonded system. There can be a true average occupation of the two minima by the same hydrogen/deuterium atom, but at two different atomic distances. This is illustrated in Fig. 6.32b. The above examples demonstrate the unique advantage of neutron diffraction. Neutron diffraction is also a powerful tool in deciphering the isoelectronic ions which is not possible by XRD. For example, in the albite ($NaAlSi_3O_8$), Na^+, Al^{3+}, Si^{4+} and O^{2-} are isoelectronic and hence have the same atomic scattering factors. However, their scattering length $\overline{b}*$ in neutron diffraction is very different. Here, the positions in the lattice can be precisely located. Another example is of spinel, $MgAl_2O_4$, wherein Mg^{2+} Al^{3+} and O^{2-} have the same electronic configurations. Neutron diffraction studies were able to clearly show that Mg^{2+} is located in tetrahedral voids, whereas Al^{3+} is in the octahedral voids. The studies on spinels have also clearly shown the existence of direct- and inverse-spinel structures. In ferrites (MFe_2O_4), the divalent and trivalent ions interchange the tetrahedral and octahedral positions. In the controversial situations, such as the positional disorder of ions, neutron diffraction is of immense help.

Bibliography

F C Phillips: *An Introduction to Crystallography* (1963) Longmans, Green & Co Ltd, London.

M J Buerger: *Elementary Crystallography* (1956) J Wiley & Sons Inc, New York.

W F De Jong: *General Crystallography* (1956) Freeman, San Franscisco.

P Terpstra: *A Thousand and One Questions on Crystallography Problems* (1952) J B Walters, Groningen.

F D Bloss: *Crystallography and Crystal Chemistry* (1971) Holt, Reinhart & Winston Inc, New York.

P Terpstra and L W Codd: *Crystallometry* (1961) Longmans, Green & Co Ltd, London.

A E H Tutton: *Crystallography* (1922) Macmillan, New York.

H E Buckley: *Crystal Growth* (1951) J Wiley & Sons Inc, New York.

H Hilton: *Mathematical Crystallography* (1963) Dover Publications, New York.

D McKie and C McKie: *Crystalline Solids* (1974) Nelson & Sons Ltd, London.

L V Azároff: *Introduction to Solids* (1960) McGraw–Hill, New York.

C W Bunn: *Chemical Crystallography* (1945) Oxford University Press, London.

R C Evans: *An Introduction to Crystal Chemistry* (1939) Cambridge University Press, London.

A F Wells: *Structural Inorganic Chemistry* (1950) Clarendon Press, Oxford.

R H Dorenus, B W Roberts and D Turnbull: *Growth and Perfection of Crystals* (1950), J Wiley & Sons Inc, New York.

L Pauling: *The Nature of Chemical Bonds* (1948) (2nd ed) Cornell University Press, Ithaca, New York.

L H Ahrens: The Use of Ionization Potentials: Part 1: Ionic radii of the elements, *Geochimica et Cosmo–Chimica Acta* **2**, 155–169 (1952).

O K Rice: *Electronic Structure and Chemical Bonding* (1940) McGraw–Hill Inc, New York.

W Hume–Rothery and G V Raynor: *The Structure of Metals and Alloys* (1956) Institute of Metals, London.

W Hume–Rothery: *Atomic Theory for Students of Metallurgy* (1946) Institute of Metals, London.

W E Addison: *Structural Principles in Inorganic Compounds* (1961) J Wiley & Sons Inc, New York.

R D Shannon and C T Prewitt: Effective Ionic Radii in Oxides and Fluorides. *Acta Crystallographia*, **38**, 925–946 (1969).

W L Bragg, G F Claringbull and W H Taylor: *Crystal Structure of Minerals* (1965) Cornell University Press, Ithaca, New York.

N N Greenwood: *Ionic Crystals, Lattice Defects and Nonstoichiometry* (1968) Butterworths, London.

F A Kröger: *The Chemistry of Imperfect Crystals* (1964) North Holland, Amsterdam.

W Van Gool: *Principles of Defect Chemistry of Crystalline Solids* (1966) Academic Press, New York.

F S Galasso: *Structures and Properties of Inorganic Solids* (1970) Pergamon Press, London.

J P Frankel: *Principles of the Properties of Materials* (1957) McGraw–Hill Inc, New York.

W T Read Jr: *Dislocations in Crystals* (1953) McGraw–Hill Inc, New York.

A H Cottrell: *Dislocations and Plastic Flow in Crystals* (1956) Oxford University Press, London.

W A Wooster: *A Textbook on Crystal Physics* (1938) Cambridge University Press, London.

A R Verma: *Crystal Growth and Dislocations* (1953), Butterworths Publications Ltd, London.

R Smoluchowski, J E Mayer and W A Weyl: *Phase Transformations in Solids* (1951) J Wiley & Sons Inc, New York.

F Seitz: *The Physics of Metals* (1953) McGraw–Hill Inc, New York.

A J Dekker: *Solid State Physics* (1957) J Wiley & Sons Inc, New York.

C Kittel: *Introduction to Solid State Physics* (1986) (6th Edn.) J Wiley & Sons Inc, New York.

A R Verma and P Krishna: *Polymorphism and Polytypism in Crystals* (1969) J Wiley & Sons Inc, New York.

S Bhagavantam: *Crystal Symmetry and Physical Properties* (1966) Academic Press, New York.

H D Megaw: *Ferroelectricity in Crystals* (1967) Oxford University Press, London.

W A Wooster: *A Textbook on Crystal Physics* (1949) Cambridge University Press, London.

F D Bloss: *An Introduction to the Methods of Optical Crystallography* (1961) Holt, Reinhart & Winston Inc, New York.

N H Hartshorne and A Stuart: *Crystals and the Polarizing Microscope* (1960) Edward and Arnold, London.

A C Hardy and F H Perrin: *The Principles of Optics* (1931) McGraw–Hill Inc, New York.

J Zussman, *Physical Methods in Mineralogy* (1967) Academic Press, London.

J Zussman (Ed): *Physical Methods in Determinative Mineralogy* (1967) Academic Press, New York.

M J Buerger: *X-ray Crystallography* (1942) J Wiley & Sons Inc, New York.

L V Azároff and M J Buerger: *The Powder Method* (1958) J Wiley & Sons Inc, New York.

L V Azároff: *Elements of X-ray Crystallography* (1968) McGraw–Hill Inc, New York.

E W Nuffield: *X-ray Diffraction Method* (1966) J Wiley & Sons Inc., New York.

H P Klug and L E Alexander: *X-ray Diffraction Procedures* (1954) J Wiley & Sons Inc, New York.

M M Woolfon: *An Introduction to X-ray Crystallography* (1970) Cambridge University Press, London.

G E Bacon: *Neutron Diffraction* (1962) Oxford University Press, London.

D B Hirsch, A Howie, R B Nicholson, D W Pashley and M J Whelan: *Electron Microscopy of Thin Crystals* (1965) Butterworths Publishers, London.

J W Cowley (Ed): *Electron Diffraction Techniques* (1993) Oxford University Press, London.

Walter Borchard–Off : *Crystallography* (translated from the German) (1993) Springer–Verlag Berlin, Heidelberg, New York.

Subject Index